Late Ordovician Articulate Brachiopods from the Red River and Stony Mountain Formations, Southern Manitoba

NRC Monograph Publishing Program

Inquiries: Monograph Publishing Program, NRC Research Press, National Research Council of Canada, Ottawa, Ontario K1A 0R6, Canada. Web site: www.monographs.nrc.ca

Correct citation for this publication: Jin, J., and Zhan, R.-b. 2001. *Late Ordovician Articulate Brachiopods from the Red River and Stony Mountain Formations, Southern Manitoba.* NRC Research Press, Ottawa, Ontario, Canada. 117 pp.

A Publication of the
National Research Council of Canada
Monograph Publishing Program

Late Ordovician Articulate Brachiopods from the Red River and Stony Mountain Formations, Southern Manitoba

Jisuo Jin
Department of Earth Sciences
University of Western Ontario
London, Ontario N6A 5B7

and

Ren-bin Zhan
Nanjing Institute of Geology and Palaeontology
Academia Sinica
Nanjing 210008, China

NRC·CNRC
NRC Research Press
Ottawa 2001

Printed in Canada on acid-free paper. ♾

ISBN 0-660-18283-1
NRC No. 44455

Canadian Cataloguing in Publication Data

Jin, Jisuo, 1957–
Late Ordovician Articulate Brachiopods from the Red River and Stony Mountain Formations, Southern Manitoba / Jisuo Jin, Ren-bin Zhan

Issued by National Research Council of Canada
ISBN-0-660-18283-1
Includes bibliographical references

1. Brachiopoda, Fossil—Manitoba.
2. Paleontology—Manitoba—Ordovician.
I. Zhan, Ren-bin
II. National Research Council Canada.
III. Title.

QE796.J56 2001 564'.68'097127 C00-901308-3

Contents

Abstract

The Upper Ordovician (Maysvillian–Richmondian) Red River and Stony Mountain formations of the northeastern Williston Basin, southern Manitoba, contain a suite of diverse articulate brachiopods, including 16 genera and 22 species. One genus, *Nasutimena*, and two species, *Holtedahlina paraprostrata* and *Parastrophinella cirrita,* are new. Brachiopods in the Dog Head and Cat Head members of the Red River Formation have a low diversity, with relatively small- to moderate-sized shells of *Diceromyonia*, *Thaerodonta*, and *Strophomena* predominant. In the overlying Selkirk Member (Maysvillian), the brachiopods become abundant and diverse, and typically have large to gigantic shells, as are characteristic of the epicontinental shelly benthos of the North American Late Ordovician equatorial fauna. The strophomenid shells are especially large and associated with abundant receptaculitid algae, large colonies of tabulate corals, colonial rugosans, stromatoporoids, gigantic gastropods and nautiloids, and ubiquitous *Thalassinoides* traces. This association points to a relatively shallow, clear, open, tropical sea with a well-oxygenated substrate. In contrast, the equally abundant brachiopods of the overlying Stony Mountain Formation (Richmondian) are dominated by relatively large orthid and rhynchonellid shells — typical elements of the *Hiscobeccus* fauna that flourished in, and was largely confined to, North America. Although the taxonomic composition of the Stony Mountain brachiopods shows a relatively high level of continuity with that of the Red River Formation, the absence of gigantic strophomenids and receptaculitids and the replacement of hermatypic corals by ahermatypic solitary rugose corals indicate that a turbid-water, muddy-bottom environment prevailed when deposits of the lower Stony Mountain Formation accumulated. The two broad Late Ordovician brachiopod biozones established for the Hudson Bay Basin, the *Tetraphalerella churchillensis – Kjaerina hartae* Biozone and the *Dinorthis occidentalis – Hiscobeccus capax* Biozone, can be recognized in the upper Red River and the lower Stony Mountain formations, respectively.

The Late Ordovician brachiopod fauna of southern Manitoba has its closest affinity to the coeval faunas in the Bad Cache Rapids and Churchill River groups of the Hudson Bay Basin, northern Manitoba, and in the Bighorn Formation of the southern Williston Basin, Wyoming. Pronounced provincialism of North American brachiopods during Ashgill time is further demonstrated by the extremely low affinity indices between the Red River fauna and the coeval brachiopod faunas of Siberia, Kazakhstan, and South China.

Résumé

Les formations de Red River et de Stony Mountain, Ordovicien supérieur (Maysville–Richmond), du nord-est du bassin Williston, au sud du Manitoba, comprennent une suite de divers Brachiopodes articulés, comprenant 16 genres et 22 espèces. Un genre, *Nasutimena*, et deux espèces, *Holtedahlina paraprostrata* et *Parastrophinella cirrita*, sont nouveaux. Les Brachiopodes des membres de Dog Head et de Cat Head de la Formation de Red River sont peu diversifiés et des coquillages de dimensions petites à modérées de *Diceromyonia*, *Thaerodonta* et *Strophomena* prédominent. Dans le membre sus-jacent Selkirk (Maysville), les Brachiopodes deviennent abondants et diversifiés et leurs coquillages sont typiquement de grands à gigantesques, comme c'est caractéristique pour les benthos épicontinentaux à coquillages de la faune nord-américaine équatoriale de l'Ordovicien tardif. Les coquillages Strophomenidés sont spécialement grands et associés à une abondance d'algues Receptaculitidés, de grandes colonies de coraux tabulés, de Rugeux coloniaux, de Stromatoporoïdés, de Gastéropodes et de Nautiloïdés géants, et de traces omniprésentes de *Thalassinoides*. Cette association indique une mer tropicale, ouverte, libre et

relativement peu profonde avec un substrat bien oxygéné. En contraste, les Brachiopodes tout aussi abondants de la Formation de Stony Mountain (Richmond) sus-jacente sont dominés par de relativement grands coquillages Orthis et Rhynchonella — éléments typiques de la faune *Hiscobeccus* qui florissait en Amérique du Nord et qui y était généralement confinée. Bien que la composition taxonomique des Brachiopodes de Stony Mountain montre un degré de continuité relativement élevé avec celle de la Formation Red River, l'absence de Strophomenidés et de Receptaculitidés géants et le remplacement de coraux hermatypiques par des Tétracorallinaires solitaires ahermatypiques indiquent que lors de l'accumulation des dépôts de la Formation de Stony Mountain, le fond présentait un environnement boueux d'eau turbide. Les deux grandes biozones de Brachiopodes de l'Ordovicien tardif établies pour le bassin de la Baie d'Hudson, c'est-à-dire la biozone *Tetraphalerella churchillensis* – *Kjaerina hartae* et la biozone *Dinorthis occidentalis* – *Hiscobeccus capax*, peuvent être reconnues respectivement dans la Formation de Red River supérieure et dans la Formation de Stony Mountain inférieure.

La plus proche affinité de la faune Brachiopode de l'Ordovicien tardif du sud du Manitoba est représentée par les faunes contemporaines des groupes de Bad Cache Rapids et de Churchill River du bassin de la Baie d'Hudson, au nord du Manitoba, et dans la Formation Bighorn du bassin Williston sud, au Wyoming. Le provincialisme prononcé des Brachiopodes de l'Amérique du Nord au cours de l'Ashgillien est aussi démontré par les indices d'affinité extrêmement bas entre la faune de Red River et les faunes de Brachiopodes contemporaines de la Sibérie, du Kazakhstan et du sud de la Chine.

Acknowledgments

R.J. Elias and G.A. Young kindly made available the Red River and Stony Mountain brachiopod collections of the Manitoba Museum of Man and Nature and the University of Manitoba. J. Dougherty provided loans of the type specimens from the Geological Survey of Canada. W.G.E. Caldwell read an early version of the manuscript and provided insightful comments and criticism. The critical reviews of A.J. Boucot (Oregon State University) and I. Percival (Geological Survey of New South Wales) greatly helped improve the manuscript. The project was funded by the Natural Sciences and Engineering Research Council of Canada (NSERC) Research Grant (J.J.), University of Western Ontario Vice-President Research Grant (J.J.) and UWO Academic Development Fund (J.J.). R.Z. acknowledges his funding from the Nanjing Institute of Geology and Palaeontology, Chinese Academy of Sciences and the National Natural Science Foundation of China (Project No. 49672083 and No. 40002001).

Introduction

Articulate brachiopods are abundant and diverse in the upper Red River and the lower Stony Mountain formations of southern Manitoba, but there have been limited systematic studies of these fossils. On the basis of early collections made in southern Manitoba by D. Gunn in 1858, R.W. Ells in 1875, R. Bell in 1879, T.C. Weston in 1884, and D.B. Dowling in 1891, Whiteaves (1880, 1896, 1897) described the brachiopods of the northeastern Williston Basin and used them to correlate the Upper Ordovician carbonate strata in southern Manitoba with coeval rocks in other parts of North America. Whiteaves (1897) also provided a summary account of the First and Second Franklin Expeditions (1819 and 1825–27 respectively) and the involvement of J. Richardson in the collection and identification of fossils from the Ordovician rocks of the Lake Winnipeg area. Formal documentation of brachiopod species and their localities from southern Manitoba was virtually non-existent prior to Whiteaves' work, although the presence of the distinctive receptaculitid algae and the giant gastropod *Maclurites* had been used by pioneer workers to date Ordovician rocks in the area. On the basis of his detailed systematic descriptions of the brachiopods and other invertebrate fossils, Whiteaves (1897) was able to assign a Trentonian age to some of the Ordovician strata in southern Manitoba, and this biostratigraphic interpretation still partly holds true today. Two of the largest Ordovician strophomenid brachiopods, *Rafinesquina lata* and *Kjaerina hartae*, were among the pioneer collections described by Whiteaves (1896) from what is now the Selkirk Member of the Red River Formation in southern Manitoba (Jin et al. 1995). Whiteaves (1895) also did a preliminary study of the brachiopods and other fossils from what is now the Stony Mountain Formation, but Okulitch (1943) carried out a more detailed taxonomic and stratigraphic study, providing a comprehensive faunal list for the various members of this formation and illustrating several new species. Other early paleontological works relevant to the Upper Ordovician rocks of southern Manitoba include those of Ulrich (1889) on bryozoans and ostracods, Leith (1952) on tabulate corals, Ethington and Furnish (1960) on conodonts, and Elias (1980) on eurypterids. Moreover, Elias (1981, 1982, 1983, 1985, 1991) made a series of studies of the corals from both the Red River and the Stony Mountain formations and used faunal data to interpret the ancient depositional environments of these strata in the Williston Basin. Elias et al. (1988) also undertook an integrated biostratigraphic study of the Fort Garry Member at the top of the Red River Formation based on conodonts and corals and dated the member as Richmondian in age. In their study of the Late Ordovician brachiopods of the Hudson Bay Lowlands, Jin et al. (1997) made comparisons to brachiopods from southern Manitoba and illustrated some specimens from the Red River and Stony Mountain formations.

The present study is part of a larger research project on the drastic changes in Late Ordovician – Early Silurian biodiversity in response to rapid fluctuations of global greenhouse–icehouse–greenhouse climatic episodes. The Late Ordovician Epoch was marked by one of the two greatest sea-level rises and floodings of the North American paleocontinent in Phanerozoic time, accompanied by rapid diversification of invertebrate faunas in shallow, tropical, epicontinental seas. Toward the end of the Late Ordovician, continental glaciation in the Gondwana landmass caused a major sea-level drawdown and marine regression from North America, bringing about one of the five major mass extinctions recorded in life history. Brachiopods were one of the most abundant and diverse groups of invertebrate animals during that time, and doubtless their diversity in the inland seas would have been particularly sensitive to eustatic sea-level fluctuations caused by the growth and decay of the Gondwana ice cap. A comparative study of the changes in brachiopod biodiversity between continental-margin basins and inland basins of Canada promises to reveal the duration, extent, intensity, and timing of the climatic changes and their effects on marine environments far from the site of the glaciation. A series of monographic studies of Late Ordovician – Early Silurian brachiopods has been completed for the Anticosti Basin, Hudson Bay Basin, and west-marginal shelves of Canada (Jin 1989; Jin et al. 1989, 1993, 1997; Jin and Lenz 1992; Jin and Chatterton 1997; Jin and Norford 1996; Dewing 1999;

Jin and Copper 2000). The Williston Basin is one of the largest inland basins of North America, and its Ordovician brachiopods constitute the subject of this study. Because of their generally great abundance and high diversity, brachiopods from the Red River and Stony Mountain formations of southern Manitoba are important to understanding the pattern, process, and rate of evolution, radiation, and extinction of the North American epicontinental marine shelly benthos during the Late Ordovician Epoch (Jin 1999, 2001). In this study, a detailed taxonomic treatment of these brachiopods is provided, coupled with an assessment of their implications for biostratigraphy, paleoecology, and paleobiogeography at continental and global scales.

The present work is based on fossil collections from the Geological Survey of Canada (Sinclair collection), the University of Manitoba, the Manitoba Museum of Man and Nature, and recent collections made by J. Jin. Early collections were made from a combination of limestone quarries and railway cuts, but recent collections were made almost exclusively from quarries, particularly the Garson Quarry (where most of the famous "Tyndall Stone" was produced for many prominent buildings in Canada), the City of Winnipeg Quarry, and the Stony Mountain Penitentiary Quarry. Most of the brachiopod shells are calcareous, with varying degrees of recrystallization and sporadic dolomitization. In general, however, the shells remain suitable for serial sectioning to reveal internal structures, and this is important because naturally exposed interiors of disarticulated shells are relatively rare for many taxa. Serial acetate peel sections were prepared by grinding the shells with a Croft Parallel Grinder and taking Transilwrap acetate peels at 0.1–0.3 mm intervals. To produce half-tone micrographs of the serial sections, the peels were scanned directly on a Zeiss petrographic microscope using a Leaf MicroLumina digital scanning camera. The digital files then were edited and printed with an Epson Stylus Pro 5000 printer. Complete shells illustrated in the present paper were photographed conventionally by coating the specimens with either ammonium chloride or magnesium oxide. The negatives were scanned using a Nikon CoolScan 2000, and the acquired digital files were edited and compiled into plates in digital format. The primary digital images generally did not receive any touch-ups except for adjustment of contrast and brightness and occasional repairs of obvious scratch marks.

Geological setting

As one of the largest intracratonic basins of North America, the Williston Basin (Fig. 1) was best delimited as an active sedimentary basin during Late Ordovician – Early Silurian times, when the Tippecanoe cratonic sedimentary sequence was deposited in western North America (Sloss 1963, 1988). During the Devonian and Mississippian periods, the Williston Basin became part of the giant Elk Point Basin, stretching from the southern Northwest Territories to North Dakota and from Alberta to Manitoba. In Late Ordovician time, however, the Williston Basin was a subcircular structure covering southwestern Manitoba, southern Saskatchewan, southeastern Alberta, North Dakota, and parts of South Dakota, Montana, and Wyoming, with the depocentre located between Williston and Bismark, North Dakota (Norford et al. 1994; Longman and Haidl 1996).

The Lower Paleozoic sedimentary succession in the Williston Basin begins with the Middle–Late Cambrian Deadwood Formation, which comprises mainly siliciclastic sandstones (mature quartz sandstones and glauconitic sandstones), micaceous siltstones, and shales (Butler et al. 1955; Hendricks et al. 1998). Sediments of the Deadwood Formation accumulated in much of the basin, although subsequent pre-Late Ordovician erosion has removed the deposits from the northeastern part of the basin (i.e., north-central Saskatchewan and much of western Manitoba). As a result, the Late Ordovician Winnipeg Formation, consisting mainly of mature quartz sandstones, silty bioturbated sandstones, and shales, lies disconformably on the Precambrian basement in southern Manitoba (the study area). The Williston Basin may have assumed its incipient shape as an intracratonic basin in Cambrian time, as the Deadwood Formation attains its greatest thickness in the Williston area, near the basin's early Paleozoic depocentre (Hein and Nowlan 1998).

Fig. 1. Isopach map of the Upper Ordovician Red River Formation in the Williston Basin, showing the major confining arches around the basin, the depositional centre between Williston and Bismark, and the outcrop belt in Manitoba (shaded area). Isopach intervals are in 100 m (adapted from Norford et al. 1994 and Longman and Haidl 1996).

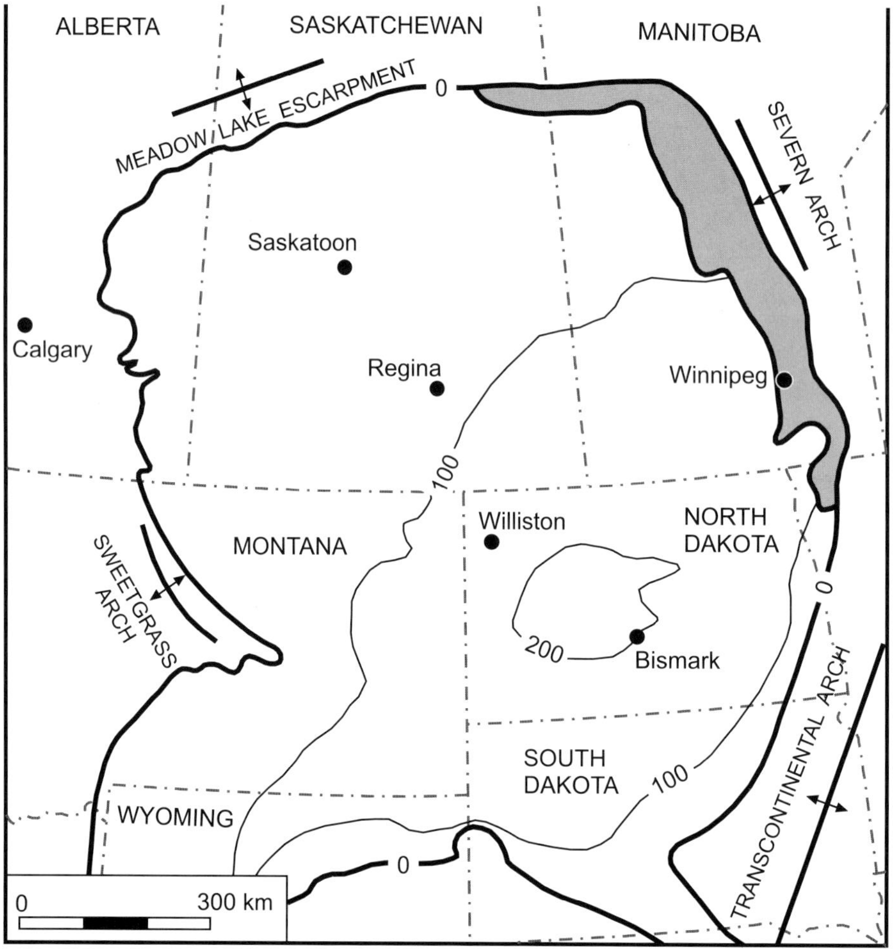

The Winnipeg Formation (Figs. 2, 3), or the Winnipeg Group of some authors (e.g., Carlson and Thompson 1987), consists of large sandstone ridges (kilometers to tens of kilometers in length or width and up to 65 m thick) in the subsurface of southern Manitoba and northeastern North Dakota (McCabe 1971; Kessler 1991). Towards the basinal depocentre, however, the lithology becomes finer grained and grades into mudstones. The Winnipeg Formation covers a large area (mostly subsurface), encompassing southern Saskatchewan, southern Manitoba, eastern Montana, North Dakota, northwestern South Dakota, northeastern Wyoming, and northwestern Minnesota. It marks the basal deposit of the Tippecanoe marine transgression (Fig. 3) and clearly delimits the subcircular, dish-shaped configuration of the Williston Basin by its concentric isopachs, which shows the greatest thickness (up to 100 m) in northwestern North Dakota (Sweet 1982). In the study area, the Winnipeg Formation is largely a mature quartz arenite from which only microfossils have been recovered previously (Oberg 1966; Sweet 1982). Examination of hand specimens (Manitoba Museum of Man and Nature MMMN 1534) from the southern Manitoba outcrop belt (Fig. 2B), however, indicates that even the coarse-grained, mature-sandstone facies of the Winnipeg Formation may be fossiliferous locally, containing such megafossils as brachiopods (*Glyptorthis* sp., *Scaphorthis* sp., and *Strophomena* sp.), crinoid ossicles, and trepostome and fenestrate bryozoans. The state of their preservation, however, is not adequate for formal description. Conodont data from Manitoba, North Dakota, and South Dakota (Oberg 1966; Sweet 1979, 1982) suggest a Trentonian (Rocklandian–Kirkfieldian) age for the Winnipeg Formation. This implies that the Winnipeg

Fig. 2. A, Map of the northeastern Williston Basin and the southwestern Hudson Bay Basin. **B**, Enlarged view of the small rectangle in the Winnipeg area, showing the Upper Ordovician formations in the outcrop belt and some classic fossil localities. Both the Penitentiary Quarry and the City of Winnipeg Quarry are within the area marked Stony Mountain (adapted from Jin et al. 1995, 1999).

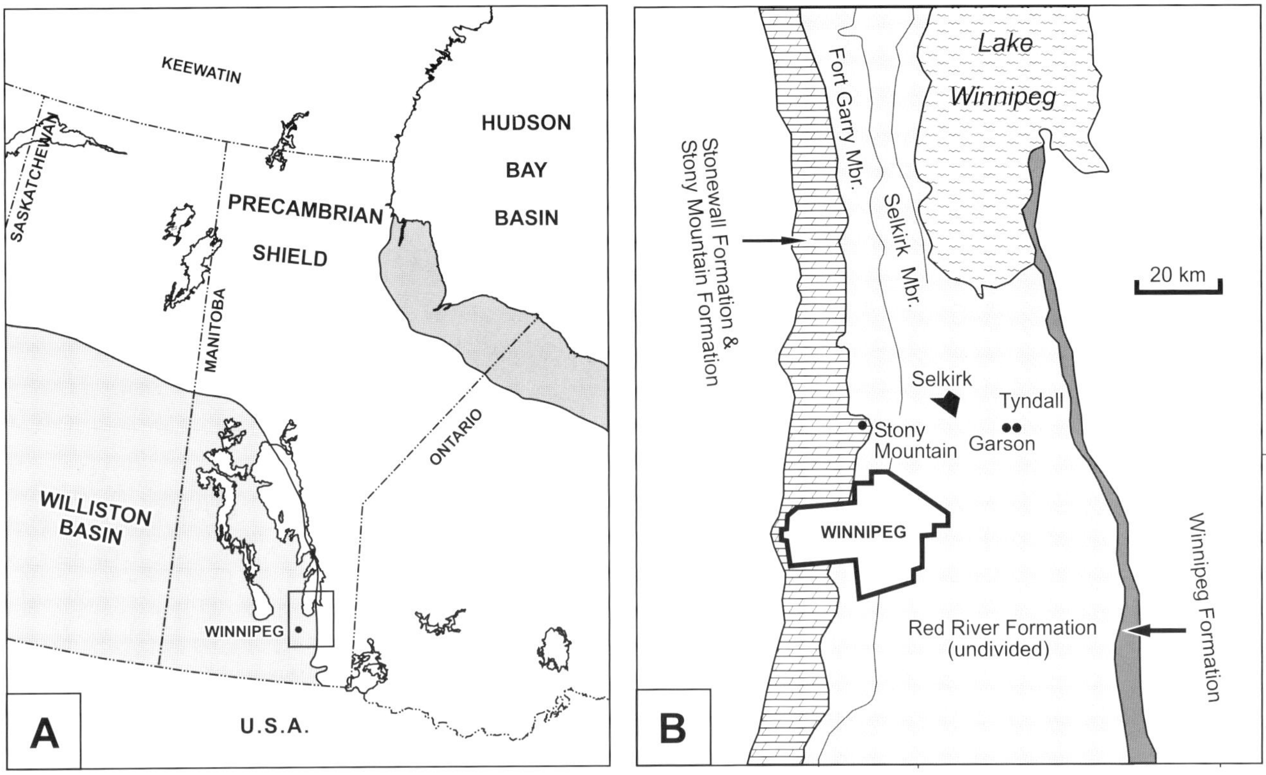

siliciclastic sediments were largely correlative to the carbonate sediments of the Advance Formation in the Rocky Mountains (Jin and Norford 1996).

In southern Manitoba, the Red River Formation is divided into the Dog Head, Cat Head, Selkirk, and Fort Garry members (Fig. 3; Dowling 1895; Foerste 1929; McCabe and Bannatyne 1970; McCabe 1971; Cowan 1971; McCabe and Barchyn 1982; Elias et al. 1988). The total thickness of the formation shows a consistent thinning from southwestern Manitoba northward. Some localized abnormal thickenings (to about 150 m) near the U.S. border are probably related to the underlying Winnipeg sand ridges (Kessler 1991). The formation is somewhat thicker (165–213 m) near the centre of the Williston Basin from southeastern Saskatchewan to the North Dakota – South Dakota border (Longman and Haidl 1996). Three or four brining-upward depositional cycles have been recognized in the formation, particularly around the basin centre (Canter 1998; Longman et al. 1998). Each of the cycles (named D, C, B, A for the U.S. portion) typically consists of bioturbated mudstone, succeeded by laminated mudstone, and by bedded anhydrite. The microbialitic (or kukersitic) laminated mudstones have been regarded as possible source rocks for the hydrocarbon reservoirs in the Red River Formation. In southern Saskatchewan, two of such anhydrite units are known in the Herald Formation (Longman and Haidl 1996), which is approximately correlative to the Fort Garry Member (Norford et al. 1994). In the study area, however, no anhydrite beds have been observed in the Red River Formation. This indicates that the dish-shaped Williston Basin was periodically reduced to a salina at the basin centre, with subaerial conditions prevailing toward the basin margin.

In southern Manitoba, the Dog Head, Cat Head, and Selkirk members constitute the open-marine stage of Cycle C, represented by the "C burrowed" member in North Dakota and the Herald Formation in Saskatchewan (Fig. 3).

The Dog Head and Cat Head members consist mainly of relatively dense, argillaceous, bioturbated mudstone, which is largely calcareous with some degree of dolomitization in the Winnipeg – southern

Fig. 3. Stratigraphic nomenclature of Upper Ordovician strata in the Williston Basin. The stratigraphic divisions are based on outcrop and subsurface sections in southern Manitoba but mainly on subsurface data in North Dakota and Saskatchewan (modified from Norford et al. 1994 and Longman and Haidl 1996).

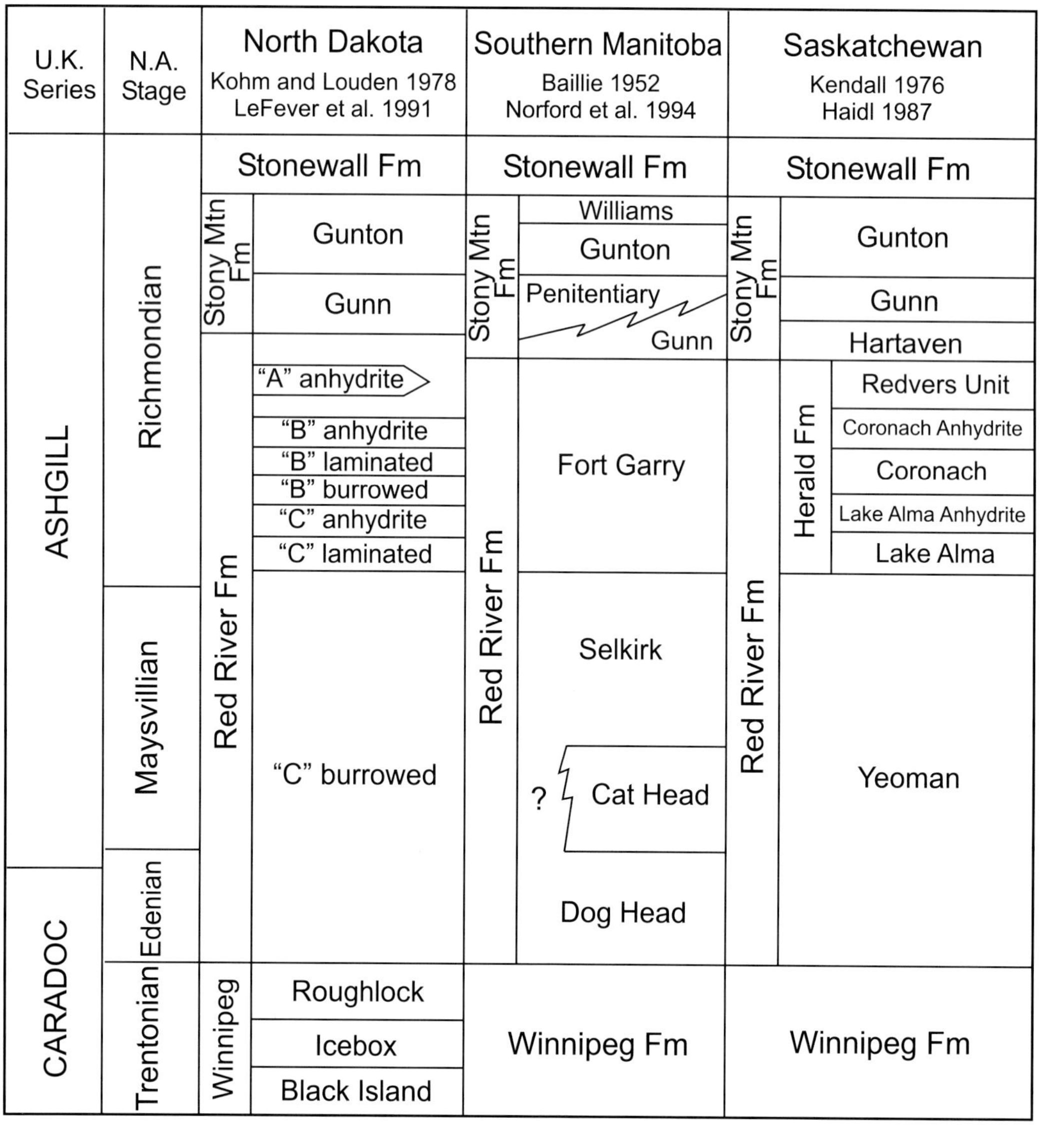

Lake Winnipeg area but completely dolomitized in south-central Manitoba, toward the margin of the Williston Basin (McCabe 1971). Brachiopods in these lower two members are characteristically low in diversity and abundance, with relatively small-shelled *Diceromyonia storeya* (Okulitch, 1943), *Thaerodonta clarksvillensis* (Foerste, 1912), and *Nasutimena fluctuosa* (Billings, 1860) being most common. Interestingly, small- and thin-shelled brachiopods (e.g., *T. clarksvillensis*) are most abundant in the Dog Head Member but become increasingly scarce up-section in the Cat Head Member and even more so in the Selkirk Member, where large and gigantic forms predominate. A similarly low diversity and general scarcity are also observed in corals from the Dog Head and Cat Head members (Elias 1991). This indicates that the carbonate sediments of these two members may have accumulated in relatively deep waters during the early stages of the Tippecanoe transgression, when the rate of sediment accumulation lagged behind the rate of sea-level rise.

The overlying Selkirk Member is marked by an abrupt increase in species diversity as well as in overall number of individuals. Nearly all brachiopod species described in the present paper are present in the Selkirk Member. In the Selkirk–Garson–Tyndall area (Fig. 2B), the Selkirk strata lie immediately below a

thin layer of glacial overburden. Gigantism is a distinct feature of the brachiopod fauna from this member. Some shells of *Oepikina lata* (Whiteaves, 1896), *Kjaerina hartae* Jin, Caldwell, and Norford, 1995, and *Tetraphalerella churchillensis* Jin, Caldwell, and Norford, 1997, are the largest Ordovician strophomenids ever recorded. In addition, the Selkirk Member also contains a varied suite of other fossils, including receptaculitid algae, stromatoporoid, corals, gastropods, cephalopods, trilobites, and ubiquitous trace fossils. This diverse biota was noted already in Dowling's (1900) extensive fossil list for the Upper Mottled Limestone (= Selkirk Member). Toward the centre of the basin, in southern Saskatchewan (Well 7-28-4-4W2), the equivalent of the Selkirk Member is the upper Yeoman Formation, which contains a 3-m-thick unit of reefal, stromatoporoid-thrombolitic boundstone (Pratt et al. 1996; Haidl et al. 1997). The biota, together with more varied lithologies of mudstone, wackestone, and localized packstone and grainstone pockets, points to a generally shallow, warm, open, well-circulated, epicontinental sea.

The Fort Garry Member at the top of the Red River Formation consists of approximately 35 m of thickly bedded, coarsely mottled dolomudstones (Fig. 2B). It is largely correlative to the Herald Formation in Saskatchewan, and the upper evaporitic beds of Member C upward through Member A in North Dakota, South Dakota, and Montana (Fig. 3), representing deposits of a semi-restricted to restricted basin. Megafossils are generally sparse, low in diversity, and rather poorly preserved (due to dissolution, dolomitization, or silicification), but include green algae, stromatoporoids, brachiopods, bryozoans, trilobites, ostracods, and crinoids. In the collections used in this study, only one species of brachiopod (*Holtedahlina paraprostrata* n. sp.) is sufficiently well preserved to allow systematic description. In contrast, conodonts are abundant and relatively diverse. On the basis of the complete biota and the lithological evidence, Elias et al. (1988) interpreted the Fort Garry Member to be composed of cycles of basin-wide synchronous brining-upward (not progradational shallowing-upward) sequences and assigned a mid-Richmondian age to the type section of the Fort Garry Member.

In the Winnipeg outcrop area, the Stony Mountain Formation is divided, in ascending order, into the Gunn, Penitentiary, Gunton, and Williams members (Fig. 3; Dowling 1900; Okulitch 1943; Baillie 1952; Sinclair and Leith 1958; Sinclair 1959; Smith 1963; Elias 1983), with a total thickness of 40 m (McCabe 1971). The formation thins northward to about 30 m (Cowan 1971) near the northern tip of Lake Winnipeg, and thickens southward to the U.S. border and westward to the Saskatchewan border to reach a maximum of 48 m. The Gunn Member comprises reddish-weathering, argillaceous wackestones and packstones, which are rich in brachiopods and corals. Many shaly beds in the member weather recessively, exposing well-preserved shelly fossils. Thousands of specimens from the member have been examined during the course of this study. The overlying Penitentiary Member is similar to the Gunn Member in lithology and fossil content, except that it is dolomitized and usually weathers yellowish tan in colour. The fossils, especially brachiopods and corals, have been preferentially dissolved, leaving large vugs in the outcrops. The dense, finely crystalline character of the Penitentiary dolomite favours preservation of exquisite internal moulds of brachiopods (Pl. 2, figs. 1–6, 12, 13; Pl. 20, figs. 1–13), which can be collected in great numbers at some localities in the Winnipeg area. Bioturbation and relatively well-preserved burrows are common in both the Gunn and the Penitentiary members. The Gunton Member typically consists of massive, poorly fossiliferous, sublithographic dolomudstone, with a few thinner, marly interbeds, but the upper part may be vuggy as a result of fossil dissolution. The Williams Member consists of arenaceous to argillaceous dolomudstone and is poorly fossiliferous.

Lithological and faunal data suggest that the Gunn and Penitentiary members were deposited in a relatively deep-water (probably just below storm wave base), low-energy, open-marine environment. This is supported by the moderately diverse, extremely abundant, well-preserved shelly benthos and abundant burrowing organisms in mud-rich sediments. The Gunton and Williams members represent a restricted (or at least periodically restricted) depositional environment. This may have been caused by the early stage of marine regression related to the onset of the Gondwana glaciation.

The overlying Stonewall Formation consists of dolomudstone and laminated dolomudstone, with interbeds of fossiliferous dolowackestone. The Stonewall Formation has a maximum thickness of 30.5 m in southern Manitoba and 36 m in southern Saskatchewan, both near the border with North Dakota (Norford et al. 1998). In southern Manitoba and southern Saskatchewan, the lower (Upper Ordovician) part

of the Stonewall Formation contains moderately common rugosans and small colonies of tabulate corals in places. The carbonate strata are interrupted by two to four thin, argillaceous marker beds near the basin margins and by corresponding anhydrite beds toward the basin centre (Kendall 1976). The particular marker bed known as the T-marker or upper T-marker in the upper Stonewall Formation has been shown by conodont data to coincide with the Ordovician–Silurian boundary (Norford et al. 1998; Nowlan and Haidl 1999). The lower Stonewall marker rests directly on the top of the Stony Mountain Formation, and the top Stonewall marker lies immediately below the base of the overlying Lower Silurian Fisher Branch Formation. Kendall (1976) interpreted the T-marker and other similar marker beds in the Upper Ordovician – Lower Silurian succession of the Williston Basin as basal lag deposits, denoting the beginning of depositional (transgressive) cycles. Johnson and Lescinsky (1986), however, interpreted the marker beds as deposits of peak regression and ensuing subaerial exposure, marking the end of depositional cycles. It should be noted that both the lower and upper contacts of the T-marker are generally sharp, and a significant depositional hiatus is most likely present at the bottom and the top of the marker. Thus, the T-marker does not necessarily have a genetical relationship to the carbonate packages below or above. Despite this, the wide geographic extent and consistent lithology of the T-marker indicate that it was deposited in a homogeneous environment of relatively uniform water depth (Jin et al. 1999). Although brachiopods have been reported from the Stonewall Formation (Brindle 1960), none collected from the southern Manitoba outcrops are sufficiently well preserved to allow systematic treatment.

At the base of the overlying Fisher Branch Formation, Jin et al. (1999) reported a widespread *Virgiana decussata* fauna in Manitoba and Saskatchewan and dated the basal Fisher Branch Formation as late Rhuddanian. As the youngest conodonts from the beds immediately below the Ordovician–Silurian boundary in the Stonewall Formation are probably of late Richmondian age (Nowlan and Haidl 1999), the upper Stonewall strata between the T-marker and the *Virgiana* beds could be of any age between Hirnantian and early late Rhuddanian. The dolomudstone and argillaceous marker beds in this interval were deposited under restricted marine conditions, with periods of subaerial exposure, which probably denote the Hirnantian marine regression and retreat of the epeiric seas from the intracratonic basins.

Biostratigraphy

The Upper Ordovician succession of southern Manitoba spans the Trentonian–Richmondian stages, and much work remains to be done to refine the details of regional biostratigraphy. Brachiopods from the Winnipeg Formation at the base of the succession are insufficiently well preserved to be biostratigraphically useful. Conodonts from the formation both near the basin depocentre (North and South Dakota) and toward the margin (southern Manitoba) indicate an early middle Trentonian (Rocklandian Kirkfieldian) age (Oberg 1966; Sweet 1979, 1982). The Red River and Stony Mountain formations are largely Ashgill (Maysvillian to Richmondian) in age.

The age of the lower Red River Formation (the Dog Head and Cat Head members) has not been precisely determined, but it is generally assumed that the basal part of the formation falls within the Edenian. The so-called "Red River fauna" (or "Arctic Ordovician Fauna" of Nelson 1959), characterized by a suite of generally abundant and diverse, and typically large-sized shelly organisms, is best represented in the Selkirk Member (Fig. 4). Elias (1981, 1985) recognized a *Grewingkia*-dominated solitary rugose coral assemblage in the Selkirk Member, which can be distinguished from a *Salvadorea*-dominated coral assemblage in the lower Stony Mountain Formation. Jin et al. (1997) identified two brachiopod biozones in the Upper Ordovician succession of the Hudson Bay Basin. The *Tetraphalerella churchillensis – Kjaerina hartae* Biozone in the upper Portage Chute and Surprise Creek formations was assigned a Maysvillian age and regarded as correlative of the *Kjaerina* Community of the Selkirk Member (Fig. 4). The presence of a fully developed *Hiscobeccus* Fauna

Fig. 4. Range chart of articulate brachiopods in the Red River and Stony Mountain formations of southern Manitoba. Fragmentary and poorly preserved brachiopods are present in the Winnipeg and Stonewall formations but are not plotted.

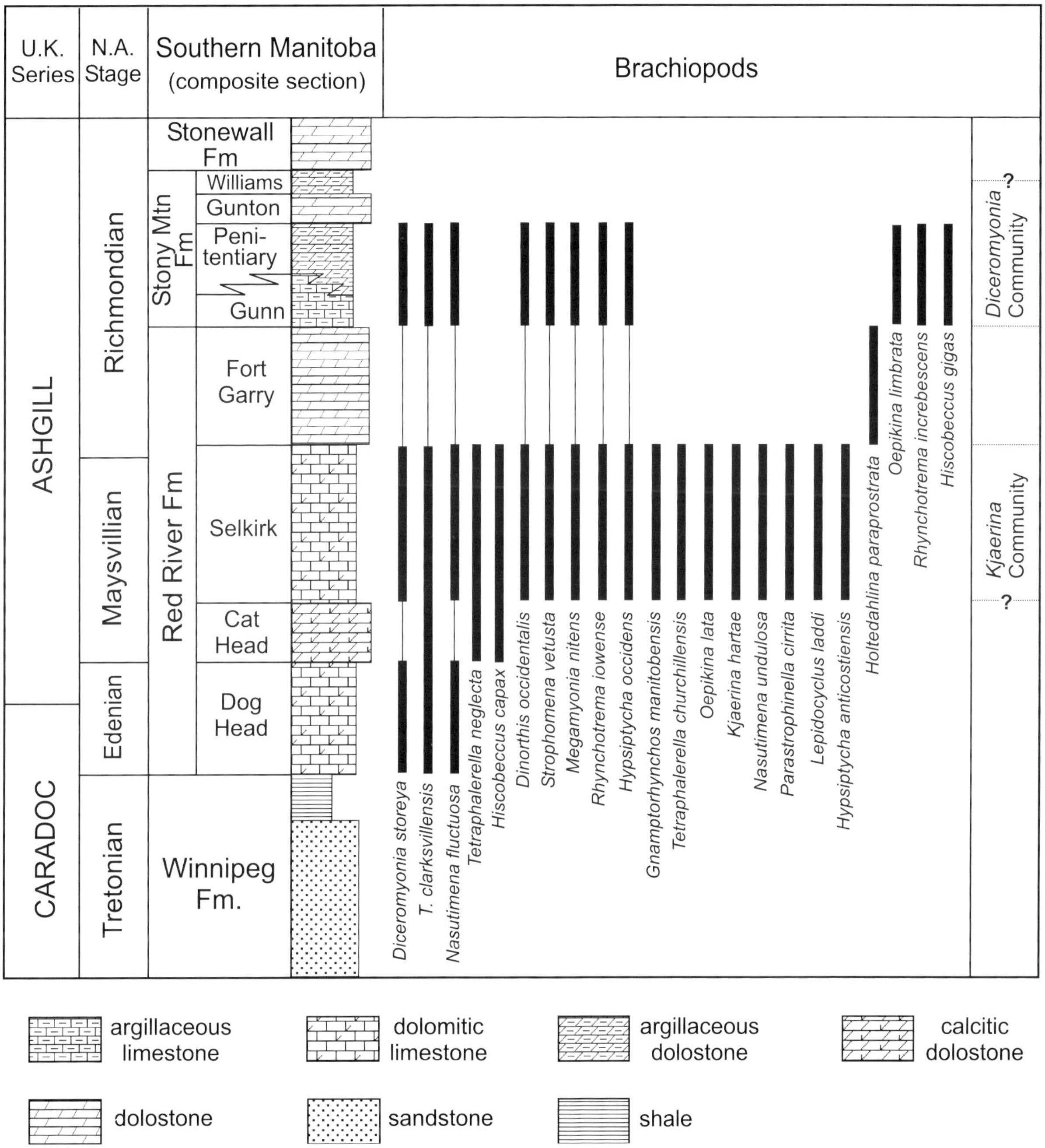

(*Hiscobeccus*, *Hypsiptycha*, and *Lepidocyclus*) indicates that the Selkirk Member is post-Edenian in age. This supports assignment of a Maysvillian age to the Selkirk Member and its correlation with the middle upper Bad Cache Rapids Group (upper Portage Chute and Surprise Creek formations).

The Stony Mountain Formation has been dated as Richmondian in age (Twenhofel et al. 1954; Nelson 1963) and can be correlated to the middle upper part of the Upper Bighorn Formation of northern Wyoming in the southern part of Williston Basin (Macomber 1970). On the basis of conodonts, Ethington and Furnish (1960) correlated the lower Stony Mountain Formation to the upper part of the Bighorn Formation of Wyoming, the upper part of the Maquoketa Formation of Iowa, and the Vauréal Formation of Anticosti Island, Quebéc. Although these authors suggested a Maysvillian to early Richmondian age for this suite of

strata, it is now widely accepted that these stratigraphic units are largely mid–late Richmondian in age. Corals, conodonts, and stromatoporoids from the underlying Fort Garry Member of the Red River Formation suggest a mid-Richmondian age (Elias et al. 1988). It thus seems reasonable to assign a late Richmondian age to the Gunn and Penitentiary members. The *Diceromyonia* Community in the lower Stony Mountain Formation (Fig. 4) shows a high degree of similarity to the *D. occidentalis* – *Hiscobeccus capax* Assemblage of the Churchill River Group of the Hudson Bay Basin (Jin et al. 1997), except for the replacement of *Hiscobeccus capax* by *H. gigas* in southern Manitoba. This indicates that the Gunn and Penitentiary members of the Stony Mountain Formation fall within the *D. occidentalis* – *Hiscobeccus capax* Biozone.

Paleoecology

Despite the number of paleontological studies of the Upper Ordovician rocks of the Williston Basin (e.g., Okulitch 1943; Ross 1957; Ethington and Furnish 1960; Le Fèvre et al. 1976; Kendall 1977; Elias 1980, 1981, 1982, 1983, 1991; Elias et al. 1988), there have been relatively few comprehensive paleoecological analyses of the Red River and Stony Mountain biotas. The brachiopods, widely regarded as useful for paleoecological and paleocommunity studies (Ziegler 1965; Ziegler et al. 1968; Boucot 1975; Potter and Boucot 1992; Boucot and Lawson 1999), have not been investigated in any detail until now (Jin 1999). Among other groups, the most relevant studies have been those of Elias (1982, 1991) on the solitary rugose coral faunas, which led to the recognition of two major transgressive–regressive marine cycles in the Red River and Stony Mountain formations.

The present taxonomic study shows that the brachiopod fauna of the Red River Formation is distinguishable from that of the Stony Mountain Formation, both in taxonomic composition and species diversity (Fig. 4; Tables 1, 2). On the basis of 12 collections from the Upper Ordovician succession, two communities are recognized: the *Kjaerina hartae* Community in the upper Red River Formation (Selkirk Member) and the *Diceromyonia storeya* Community in the lower Stony Mountain Formation (Gunn and Penitentiary members). There has been long debate about the use of "community" versus "association" in the study of fossil assemblages (e.g., Petersen 1914; Thorson 1957; Pickerill and Brenchley 1979; Lockley 1983). The concept of paleocommunity, reconstructed on the basis of fossil assemblages, is well entrenched in the literature (Boucot and Lawson 1999) and this usage is followed herein.

In statistical analysis of brachiopod communities, some species may be represented largely by disarticulated valves. In such cases, the numbers of ventral and dorsal valves are counted separately, with the greater of the two numbers being taken as a proxy for the number of individuals of a given species in the community. The statistical indices of communities follow those of Calef and Hancock (1974), with the exception of species diversity, which is calculated using a formula recommended by Ziegler et al. (1968). In describing the brachiopod communities, other important fossil groups (e.g., corals) are also considered for synecological analysis.

The *Kjaerina hartae* Community

Brachiopods in the lower Red River Formation (Dog Head and Cat Head members) are generally low in species diversity and individual abundance, and no distinct brachiopod communities can be recognized. The *Kjaerina hartae* Community, best developed in the overlying Selkirk Member (Table 1), is characterized by a relatively high species diversity, moderate abundance, and notable gigantism of some species (e.g., *Tetraphalerella churchillensis*, *Oepikina lata*, and *Kjaerina hartae*).

The *Kjaerina hartae* Community shows a high degree of faunal similarity to the *Tetraphalerella churchillensis* – *Kjaerina hartae* Assemblage of the Bad Cache Rapids Group of the Hudson Bay Basin (Jin et al. 1997). According to the Otsuka, Dice, and Fager indices recommended by Rong et al. (1995), the average similarity coefficient between the *Tetraphalerella churchillensis* – *Kjaerina hartae* Assemblage and the *Kjaerina hartae* Community is 0.35 at the specific

Table 1. Composition of the *Kjaerina hartae* Community from the Red River Formation, southern Manitoba. The total number (*N*) of specimens of a given species in a given member are divided into complete shells or moulds : ventral valves or moulds : dorsal valves or moulds.

	Dog Head $N = 58$	Cat Head $N = 15$	Selkirk $N = 114$
Dinorthis occidentalis			1:0:1 (2)
Gnamptorhynchos manitobensis			12:1:3 (15)
Diceromyonia storeya	1:1:0 (2)		6:2:2 (8)
Thaerodonta clarksvillensis	55:0:0 (55)	8:0:0 (8)	2:0:0 (2)
Strophomena vetusta			0:0:1 (1)
Nasutimena fluctuosa	0:0:1 (1)		4:0:1 (5)
Nasutimena undulosa			2:0:0 (2)
Tetraphalerella neglecta		0:2:4 (4)	0:1:2 (2)
Tetraphalerella churchillensis			0:4:2 (4)
Oepikina lata			5:12:1 (17)
Kjaerina hartae			6:13:0 (19)
Megamyonia nitens			2:4:0 (6)
Parastrophinella cirrita			2:0:0 (2)
Rhynchotrema iowense			6:0:0 (6)
Lepidocyclus laddi			1:0:0 (1)
Hypsiptycha anticostiensis			4:0:0 (4)
Hypsiptycha occidens			2:1:1 (3)
Hiscobeccus capax		1:0:0 (1)	6:1:0 (7)

level and 0.61 at the generic level. This indicates that there were frequent faunal exchanges of brachiopod species between the two large intracratonic basins — Hudson Bay and Williston — during Maysvillian time.

The Selkirk Member contains an abundant and diverse biota (Elias 1982, 1991; Jin 1999), with many forms showing gigantism. There are well over a hundred species of receptaculitid green algae (*Fisherites*), stromatoporoids (*Cystostroma*, *Beatricea*), tabulate corals (*Calapoecia*, *Catenipora*, *Manipora*, *Protrochiscolithus*, *Saffordophyllum*; see Leith 1952), rugose corals (*Palaeophyllum*, *Grewingkia*, *Deiracorallium*, *Salvadorea*, *Bighornia*, *Complexophyllum*; see Elias 1991), bryozoans, brachiopods, gastropods (*Ectomaria*, *Hormotoma*, *Loxonema*, *Maclurites*), nautiloids (*Armenoceras*, *Cyclendoceras*, *Diestoceras*, *Lambeoceras*, *Narthecoceras*, *Wilsonoceras*, *Orthoceras*), trilobites (Westrop and Ludvigsen 1983), crinoids, and abundant trace fossils (*Thalassinoides*, *Trypanites*). Several genera show distinct gigantism, as seen in the receptaculitid algae, solitary rugose corals (e.g., *Grewingkia*), brachiopods (*Oepikina lata*, *Tetraphalerella churchillensis*, and *Kjaerina hartae*), gastropods (*Maclurites*), and most nautiloids. The combined high diversity, great abundance, and gigantism of the Selkirk biota are analogous to those of modern tropical shallow marine benthos (particularly reef benthos) and most strongly support the interpretation of a shallow, open, paleoequatorial (Scotese and McKerrow 1990), epicontinental sea for the deposits of the Selkirk Member. In addition to the ubiquitous *Thalassinoides* trace fossils and common, localized shelly grainstones, the Selkirk biota can be assigned to the Benthic Assemblage 2–3 in terms of Boucot's (1975) scheme. Compared to the scheme of Ordovician brachiopod benthic assemblages of Potter and Boucot (1992), however, the Selkirk biota has a notably higher diversity than the other BA2-3 faunas of Laurentia documented by these authors.

Brachiopod shells are sparse in the overlying Fort Garry Member in which the *Kjaerina hartae* Community is totally absent and only rare, relatively small shells of *Holtedahlina* are present. Conodonts, corals, and stromatoporoids, however, are relatively common to abundant in the member and indicate a shallow subtidal depositional environment (Elias et al. 1988). Coral data (Elias 1985) show that there is a change from a *Grewingkia*-dominated assemblage in the Selkirk Member to a *Salvadorea*-dominated assemblage in the Fort Garry Member and the overlying lower Stony

Table 2. Composition of the *Diceromyonia storeya* Community from the Gunn and Penitentiary members of the Stony Mountain Formation, southern Manitoba. The total number (*N*) of specimens of a given species in a given member are divided into complete shells or moulds : ventral valves or moulds : dorsal valves or moulds.

	27185 *N* = 89	37130 *N* = 2062	JJ93-1a *N* = 125	JJ93-2 *N* = 75	JJ93-3 *N* = 177	JJ93-4 *N* = 71	27186 *N* = 42	55-60 *N* = 169
Dinorthis occidentalis	10:5:14 (24)	241:161:132 (402)	14:22:21 (36)	9:16:16 (25)	15:19:11 (34)	8:10:7 (18)	2:1:3 (5)	10:20:20 (30)
Diceromyonia storeya	13:17:13 (30)	456:152:160 (616)	2:1:1 (3)	6:8:10 (16)	32:52:41 (84)	19:10:8 (29)	1:11:11 (12)	2:28:31 (33)
Thaerodonta clarksvillensis								0:2:9 (9)
Strophomena planumbona		5:1:0 (6)						
Strophomena vetusta		3:3:3 (6)	0:1:0 (1)				0:0:1 (1)	0:0:1 (1)
Nasutimena fluctuosa		11:8:16 (27)	0:1:4 (4)		1:2:1 (3)	0:0:2 (2)	0:1:3 (3)	1:7:10 (11)
Oepikina limbrata	1:2:0 (3)	35:119:109 (154)	2:10:7 (12)	0:3:3 (3)	0:0:2 (2)	1:1:2 (3)	0:2:3 (3)	0:20:4 (20)
Megamyonia nitens	0:7:0 (7)	45:67:1 (112)				0:3:0 (3)	0:0:1 (1)	
Rhynchotrema iowense						2:0:0 (2)		
Rhynchotrema increbescens								0:0:1 (1)
Hypsiptycha occidens	5:0:2 (7)	319:6:9 (328)	38:1:0 (39)	4:0:0 (4)	0:1:0 (1)		0:1:1 (1)	1:1:1 (2)
Hiscobeccus gigas		58:11:0 (69)	2:3:0 (5)	1:0:0 (1)	2:0:0 (2)	3:2:1 (5)	1:1:0 (2)	3:9:7 (12)

Mountain Formation, which has been interpreted to record a change from an open marine, relatively high-energy environment to a partly restricted, low-energy, shallower-water environment, particularly for the Fort Garry Member. In southern Saskatchewan, North Dakota, and South Dakota, strata correlative to the Fort Garry Member, named the Herald Formation or C to A laminated dolomite and anhydrite units, contain prominent anhydrite beds (Longman and Haidl 1996). The lack of anhydritic beds in the southern Manitoba outcrop belt has been interpreted as the result of subsequent dissolution (Elias et al. 1988). The terminal deposits of the Red River Formation, therefore, represent an interval of drastic drop in relative sea level and episodic restriction of the Williston Basin to a salina or sabkha-dominated depositional setting. The open marine *Kjaerina hartae* Community, which evolved in, and was adapted to, a shallow-water, open-marine environment, evidently could not survive under such restricted conditions.

The *Diceromyonia storeya* Community

Despite the richness of brachiopods in the Stony Mountain Formation, there have been no detailed studies of their community structure or ecology, apart from some brief treatments of the faunal composition (Okulitch 1943; Jin et al. 1997). The *Diceromyonia storeya* Community occurs mainly in the Gunn and Penitentiary members of the lower Stony Mountain Formation, although some of its elements may extend as high as the overlying Stonewall Formation (Stearn 1956). In individual abundance and species diversity, the *Diceromyonia storeya* Community in the Gunn Member is remarkably similar to that in the Penitentiary Member (Fig. 5; Table 2). Among the nine large collections representing the *Diceromyonia storeya* Community, six are from the reddish-brown, argillaceous wackestone and packstone of the Gunn Member and three from the yellowish-buff, dolowackestone and dolopackstone of the Penitentiary

Fig. 5. Relative abundance of the component brachiopod species in the *Diceromyonia storeya* Community of the lower Stony Mountain Formation. Predominance of *Diceromyonia storeya* and *Dinorthis occidentalis* are clearly shown in samples from both the Gunn and the Penitentiary members.

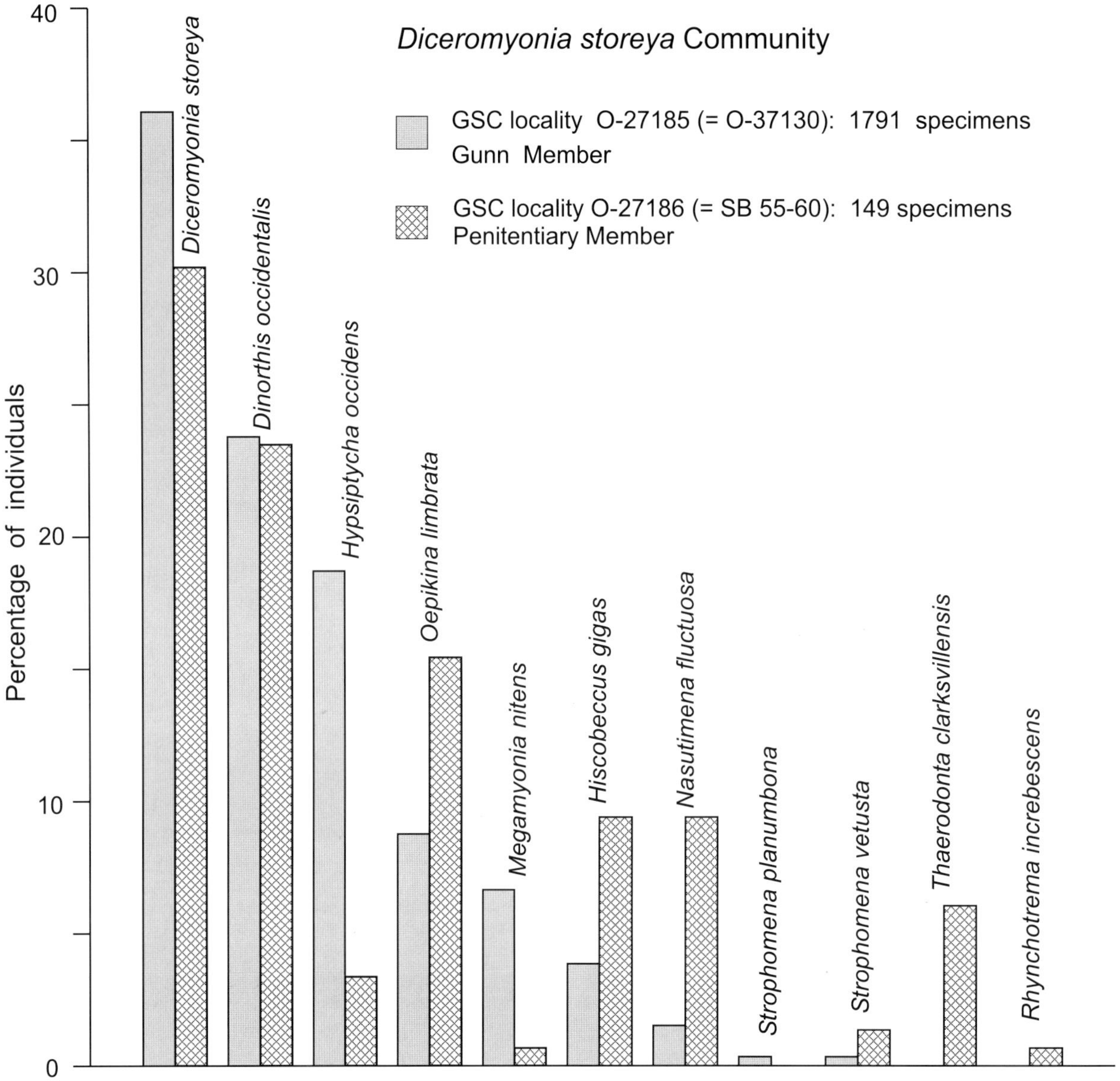

Member. Compared to the *Kjaerina hartae* Community, the *Diceromyonia storeya* Community shows a number of distinct features:

1. Low rate of species input. Seven out of 18 species from the *Kjaerina hartae* Community are found to extend into the *Diceromyonia storeya* Community, and only four species are new to the Stony Mountain Formation. This translates to a mean similarity coefficient value of 0.45. At the generic level, the two communities have a relatively high similarity coefficient of 0.72, which implies that the communities evolved as a continuous succession rather than being a replacement of one by the other (sensu Boucot 1975 and Rong 1986).
2. Predominance of a small number of species in the community (Fig. 5). The decrease in species diversity and concomitant increase in the abundance and predominance of a few taxa (e.g., *Dinorthis occidentalis*, *Diceromyonia storeya*, *Hiscobeccus gigas*) suggest that environmental stress was higher during deposition of the Stony Mountain sediments than that of the Selkirk Member. On the other hand, the Stony Mountain fauna is far from being a monospecific community, as indicated by the co-existence of a

moderately diverse and abundant coral fauna. The epicontinental sea covering the Williston Basin may have changed from being clear to turbid when the Stony Mountain deposition began. This is indicated by the higher content of siliciclastic silts and iron-rich minerals in the Gunn and Penitentiary members, probably a result of lowered sea level and increased erosion and weathering of the adjacent Precambrian rocks. This interpretation is supported by the proliferation of solitary, ahermatypic, rugose corals, which generally preferred muddy-bottom environments, and the lack or paucity of receptaculitid algae, large colonial, hermatypic corals and stromatoporoids, and giant strophomenid brachiopods. The presence of microborings in some shells from the Gunn Member can be regarded as supporting evidence for a shallow, nearshore environment (Perkins and Halsey 1971; Perkins and Tsentas 1976; Brett et al. 1993).

3. Much reduced gigantism. Most brachiopod shells in the *Diceromyonia storeya* Community are medium-sized, with the exception of *Dinorthis occidentalis* and *Hiscobeccus gigas*, which have relatively large shells for their genera. The lack of giant strophomenids characteristic of the *Kjaerina hartae* Community is striking. Some brachiopod species in the Stony Mountain Formation even show a trend toward reduced shell size (e.g., *Hypsiptycha occidens*). Some taxa of other fossil groups (e.g., rugose corals and trilobites) also show comparatively small sizes.

Both brachiopods and solitary rugose corals are characterized by drastic decreases in both diversity and abundance in the massive dolomudstones of the overlying Gunton Member and the argillaceous mudstones of the Williams Member. These strata are interpreted to have been deposited in a restricted, hypersaline environment (Elias 1982, 1991).

As indicated in Jin et al. (1997), the *Diceromyonia storeya* Community of the Stony Mountain Formation has a high degree of similarity to the brachiopods from the Churchill River Group of the Hudson Bay Lowlands. This is confirmed by the present taxonomic study, which shows that the faunal similarity coefficient of the two regions attains a value of 0.53 at the specific level and 0.66 at the generic level. Compared to the *Diceromyonia storeya* Community of southern Manitoba, the Churchill River brachiopod fauna has a higher level of diversity, with 14 genera and 22 species. These taxa, however, show a similar trend toward smaller shell sizes relative to the antecedent *Tetraphalerella churchillensis – Kjaerina hartae* fauna of the Bad Cache Rapids Group. *Dinorthis occidentalis*, *Diceromyonia storeya*, *Hypsiptycha occidens*, *Oepikina limbrata*, and *Megamyonia nitens* are diagnostic elements of the brachiopod faunas in both the Stony Mountain Formation and the Churchill River Group (Jin et al. 1997). *Rhynchotrema*, one of the most common rhynchonellid brachiopods in the Upper Ordovician rocks of North America, is unusually rare in the Williston and the Hudson Bay basins. Such paucity suggests that the genus, which initially evolved in marginal basins, preferred cooler- or deeper-water environments. This observation is further supported by the moderate proliferation of *Rhynchotrema* in the Ellis Bay Formation (Hirnantian glacial and interglacial interval) and prominent post-extinction recovery in the deep-water (estimated 80–120 m) settings of the Rhuddanian Merrimack Formation of Anticosti Island, Québec (Jin 1989).

Paleobiogeography

For the purpose of regional, and particularly global, paleobiogeographic analyses, the Red River and Stony Mountain brachiopods are regarded as a single brachiopod fauna because of the continuity from the *Kjaerina hartae* Community to the *Diceromyonia storeya* Community. In keeping with traditional usage, the term "Red River Brachiopod Fauna" is applied to the brachiopods in the Red River and Stony Mountain formations of southern Manitoba.

The Red River Brachiopod Fauna consists of 16 genera and 23 species in which the strophomenoids represent 50% and the orthoids 19% of the generic diversity. Similar proportions of major brachiopod groups are found in many other coeval brachiopod faunas, such as those of the Hudson Bay Lowland and South China (Jin et al. 1997; Zhan and Cocks 1998).

Compared to the Ashgillian brachiopod faunas of Siberia, Baltica, Kazakhstan, South China, and other paleoplates located in the Late Ordovician tropical to subtropical zones, the Red River Brachiopod Fauna shows a high degree of provincialism (see Appendix A and Table 3). This is in keeping with the general trend toward strong provincialism in the North American shelly benthos during Late Ordovician time (Boucot 1983; Sheehan and Coorough 1990; Jin 1996, 2001). Among the 15 genera described in the paper, only three (*Thaerodonta*, *Strophomena*, and *Holtedahlina*) were cosmopolitan. Most prevalent taxa, such as *Diceromyonia*, *Tetraphalerella*, *Megamyonia*, *Lepidocyclus*, *Hypsiptycha*, and *Hiscobeccus*, are largely confined to North America. *Platystrophia* and *Rhynchotrema*, two of the most common genera in Upper Ordovician rocks of North America, appear to be absent from the South China and North China plates (Zhan and Li 1998). At the species level, many brachiopods were largely restricted to the Williston and the Hudson Bay basins, particularly the giant strophomenoids (*Tetraphalerella churchillensis*, *Oepikina lata*, *Kjaerina hartae*, and *Nasutimena undulosa*), whereas other species (e.g., *Nasutimena fluctuosa*, *Strophomena vetusta*, *Oepikina limbrata*, *Hiscobeccus gigas*, *Rhynchotrema iowense*, and *Hiscobeccus capax*) were confined to the North American epeiric seas.

Table 3 shows the affinity indices (AI) between eight Ashgill (Maysvillian–Richmondian) brachiopod faunas from the following stratigraphic units and regions:

1. Red River and Stony Mountain formations of southern Manitoba;
2. Bad Cache Rapids and Churchill River groups of the Hudson Bay Lowlands (Jin et al. 1997);
3. Upper Bighorn Formation of Wyoming (Macomber 1970);
4. Maquoketa Formation of Iowa (Wang 1949);
5. Arnheim and Fernvale formations of Tennessee (Howe 1969, 1988);
6. Orlov Horizon (mid-Ashgill), Altai Mountains, Siberian Plate (Kulkov and Severgina 1989);
7. Upper Chokpar Formation and Dulankara Horizon (mid-Ashgill), Dulankara, Kazakhstan (Nikitin et al. 1980; Klenina et al. 1984);
8. Xiazhen and Changwu formations of South China (Zhan and Cocks 1998).

The affinity indices are calculated using three different formulae recommended by Rong et al. (1995):

Otsuka index: $$\mathrm{AI} = \frac{C}{\sqrt{N_1 N_2}}$$

Dice index: $$\mathrm{AI} = \frac{2C}{N_1 + N_2}$$

Fager index: $$\mathrm{AI} = \frac{C}{\sqrt{N_1 N_2}} - \frac{1}{2\sqrt{N_2}}$$

where N_1 = total number of taxa in Fauna 1, N_2 = total number of taxa in Fauna 2, and C = number of common taxa of the two faunas (when $N_2 > N_1$).

The average value of the three indices is taken to determine the degree of taxonomic similarity between any two given faunas (Table 3). The result shows that:

1. The brachiopod fauna of southern Manitoba (northeastern Williston Basin) has a higher similarity index (0.61) with that of the Hudson Bay Lowlands (Hudson Bay Basin) than its similarity index (0.56) with the brachiopod fauna of Wyoming (southern Williston Basin). This is a strong indication that the Williston and Hudson Bay basins were well connected during the Late Ordovician (Maysvillian and Richmondian) and favoured free migration of shelly benthos. The Severn Arch separating the two basins was most likely inactive and submergent at the time (see Fig. 1).
2. Among North American epicontinental basins, the Red River Brachiopod Fauna of the Williston Basin is much more closely related to that of the Hudson Bay Lowlands (0.61) and Iowa (0.55) than to any other regions considered. This indicates that the brachiopods from the Williston, Hudson Bay, and Iowa basins fall within a unitary realm of the North American epicontinental brachiopod faunal province (Red River Faunal Province) during Maysvillian and Richmondian times.
3. The Ashgill brachiopod fauna of Tennessee, although most closely related to that of Iowa, has a considerably lower affinity coefficient with the Red River Brachiopod Fauna of the Williston Basin, probably as a result of more frequent faunal exchanges with the shelly benthos of the marginal basins of North America.

Table 3. Ashgill brachiopod faunal affinity indices of eight regions in four paleoplates: SMN, southern Manitoba; HBL, Hudson Bay Lowlands; IOW, Iowa; WYM, northern Wyoming; TEN, Central Basin of Tennessee; SIB, Altai Mountains, Siberia; KAZ, Dulankara, Kazakhstan; SCH, Jiangshan, Changshan, and Yushan counties, Southeast China. Numbers in the upper right part of table are the average of the three values (Otsuka, Dice, and Fager indices) in the corresponding lower left part of the table (see text for more detailed discussion).

	SMN	HBL	WYM	IOW	TEN	SIB	KAZ	SCH
SMN	1	0.6037	0.5592	0.5496	0.3233	0.0644	0.118	0.0572
HBL	0.6455 0.6452 0.5205	1	0.6469	0.5077	0.3351	0.2174	0.1227	0.0125
WYM	0.6013 0.6000 0.4763	0.6901 0.6897 0.5610	1	0.589	0.4126	0.1738	0.1816	0.0625
IOW	0.5893 0.5882 0.4714	0.5477 0.5455 0.4299	0.6299 0.6250 0.5121	1	0.6466	0.2431	0.0619	0.0527
TEN	0.3638 0.3636 0.2425	0.3757 0.3750 0.2545	0.4537 0.4516 0.3325	0.6860 0.6857 0.5681	1	0.2506	0.0645	0.0127
SIB	0.0981 0.0952 0.0000	0.2532 0.2439 0.1551	0.2097 0.2000 0.1116	0.2774 0.2727 0.1793	0.2854 0.2791 0.1873	1	0.2074	0.3166
KAZ	0.1531 0.1500 0.0510	0.1581 0.1538 0.0561	0.2182 0.2105 0.1162	0.0962 0.0952 –0.0058	0.0990 0.0976 –0.0030	0.2402 0.2400 0.1421	1	0.4021
SCH	0.0834 0.0833 0.0000	0.0456 0.0426 –0.0427	0.0945 0.0870 0.0061	0.0833 0.0800 –0.0051	0.0429 0.0408 –0.0455	0.3467 0.3448 0.2583	0.4330 0.4286 0.3446	1

4. Globally, the Red River Brachiopod Fauna has a negligible affinity with the brachiopod faunas of Siberia, Kazakhstan, and South China. This further confirms the pronounced provincialism of the North American epicontinental brachiopod fauna during Ashgill time (Sheehan and Coorough 1990; Jin 1996, 1999, 2001).

On the basis of their brachiopod data from South China, Baltica, and Avalonia, Rong and Harper (1999) recognized a moderate brachiopod radiation in mid-Ashgill time, shortly before the first episode of the Late Ordovician mass extinction. In North America, radiation of the epicontinental shelly benthos, with brachiopods as a major group, largely took place during the Ashgill (Maysvillian and Richmondian). Despite the high level of provincialism, therefore, there appears to be a global event of brachiopod diversification during the Ashgill, both in the tropical epicontinental seas and in deeper-water shelves. Radiation of the Red River Brachiopod Fauna is largely coeval with the rapid dispersal of the relatively diverse, deeper-water *Foliomena*-bearing Fauna during the early Ashgill (Rong 1984; Rong and Zhan 1995, 1996; Cocks and Rong 1988; Rong et al. 1999) and the development of the equally diverse, shallow- and perhaps cool-water *Altaethyrella* Fauna during mid-Ashgill time in South China, Kazakhstan, and the Gorny Altai (Zhan and Rong 1995; Zhan and Cocks 1998; Zhan and Li 1998).

Systematic Paleontology

All the figured specimens are deposited in the Geological Survey of Canada (GSC), Ottawa, the Peabody Museum of Yale University (YPM), or the Manitoba Museum of Man and Nature (MMMN). Additional collections include those from GSC, MMMN, and the University of Manitoba (UM).

Abbreviations: *L*, shell length; *W*, shell width; W_1, hingeline width; W_2, width of sulcus; *T*, shell thickness (conjoined valves); *D*, depth of single valves; AVG, average; STD, standard deviation; MIN, minimum; and MAX, maximum.

Numbers of all serial sections denote distances from the shell apex along the commissural plane.

Qualitative terms used to describe shell size are based on the average length or width measured for a sample: small, <10 mm; medium, 10–20 mm; large, 20–40 mm; very large, >40 mm.

Order ORTHIDA Schuchert and Cooper, 1932

Superfamily ORTHOIDEA Woodward, 1852

Family PLAESIOMYIDAE Schuchert, 1913

Genus *Dinorthis* Hall and Clarke, 1892

Type species. *Orthis pectinella* Emmons, 1842. Trenton Limestone, New York.

Remarks. In the Treatise (Williams and Wright 1965), *Dinorthis* was regarded as a subgenus of the *Plaesiomys* group, as compared to Schuchert and Cooper's (1932) treatment of *Plaesiomys* as a subgenus of *Dinorthis*. In the revised brachiopod volumes of the Treatise on Invertebrate Paleontology, Williams and Harper (2000) treated *Dinorthis* as an independent genus, as in Jin et al. (1997). Separation of the two genera is further supported in this study on the basis the following observations:

1. *Dinorthis* typically has simple, coarse costae that rarely show bifurcation, whereas *Plaesiomys* typically has much finer costae that usually show bifurcation, commonly in the ventral valve, and intercalation, predominantly in the dorsal valve.
2. *Dinorthis* usually has a rectimarginate anterior commissure, as compared to the distinctly uniplicate anterior commissure in *Plaesiomys*.
3. In *Dinorthis*, the ventral diductor muscle scars are strongly bilobed, forming a prominent, rounded, medial notch (emargination) at the anterior margin of the muscle field; the adductor scars extend right to the antero–medial notch. In *Plaesiomys*, the ventral diductor scars tend to be scalloped especially in their lateral parts, with inconspicuous anterior bilobation and medial emargination; the diductor muscle scars are centrally located within the diductor muscle field, and the anterior margin of the adductor scars is some distance from the anterior margin of the diductor muscle field (Williams and Wright 1965, p. H320; Williams and Harper 2000, p. 748).

Notes on age and distribution. Williams and Harper (2000) recorded *Dinorthis* only from Caradoc strata of the eastern United States. This should be revised, as the genus is common through Caradoc–Ashgill strata in both marginal and intracratonic basins (especially the Williston Basin) of North America (see discussion below).

Dinorthis occidentalis (Okulitch, 1943)

Pl. 1, figs. 1–22; Pl. 2, figs. 1–13; Pl. 3, figs. 1–5; Pl. 21, fig. 2; Figs. 6–9

1943 *Pionorthis occidentalis Okulitch*, pp. 71–72, pl. 1, figs. 8–10.

1943 *Pionorthis* cf. *carletona* (Twenhofel); Okulitch, pl. 1, fig. 7.

1957 *Dinorthis* (*Plaesiomys*?) cf. *D.* (*P.*) *occidentalis* Ladd; Ross, pl. 37, figs. 16, 19, 20, 23.

1957 *Dinorthis* (*Pionorthis*?) cf. *D.* (*P.*) *occidentalis* Okulitch; Ross, pl. 37, figs. 17, 18, 21, 22.

1957 *Dinorthis* (*Pionorthis*?) n. sp. Ross, pl. 38, figs.1, 2, 5, 6.

1957 *Dinorthis* (?) sp. Ross, pl. 38, figs. 3, 4, 7, 8, 11.

1970 *Plaesiomys* (*Dinorthis*) *occidentalis* (Okulitch); Macomber, pp. 430–433, pl. 75, figs. 12–15; pl. 76, figs. 1–27.

1997 *Dinorthis occidentalis* (Okulitch); Jin et al., p. 21, pl. 1, figs. 13–16; pl. 2, figs. 1–17.

Types. The holotype, GSC 2043, is a dorsal valve, subsequently broken into two pieces after Okulitch's study; the paratype, GSC 2043a is a complete dorsal valve partly covered with rock matrix. Both specimens show rather poor preservation and were not photographed for this study. Gunn Member, Stony Mountain Formation (Richmondian), Stony Mountain.

Additional material examined. Penitentiary Member, City of Winnipeg Quarry (GSC localities O-27186, C-205933, and 55-60): 5 conjoined shells, 1 dorsal and 2 ventral valves; 23 dorsal internal and 14 external, 21 ventral internal and 14 external moulds, and 31 complete internal moulds.

Gunn Member (GSC loc. O-27185, O-37130, C-205926, C-205928, C-205930, C-205931): 297 conjoined shells, 201 dorsal and 233 ventral valves.

Selkirk Member, Garson Quarry (GSC loc. C-205935): 1 conjoined shell and 1 dorsal valve.

Description. Shell medium-sized to large, slightly transverse, subcircular to subrectangular in outline (Figs. 6, 7), with average length of 17.8 mm (maximum 28.2 mm), width 21.9 mm (maximum 32.7 mm), and thickness 9.6 mm (maximum 18.9 mm). Lateral profile ranging from nearly equibiconvex to strongly dorsibiconvex. Ventral valve of adult shells usually having low convexity in postero–medial portion, with flattened to weakly concave peripheral areas (Pl. 1, figs. 7, 14). Dorsal valve evenly convex, with shallow, inconspicuous sulcus developed in most specimens. Hingeline straight, extending for slightly over two-thirds of maximum shell width (W_1/W ratio at about 0.7) throughout ontogeny, with rounded cardinal extremities. Ventral interarea apsacline, with open delthyrium. Dorsal interarea orthocline, with minute, erect beak and open notothyrium. Shell covered by coarse, simple, rounded costae of uniform strength from apex to anterior margin. Bifurcation of costae occurring near anterior margin in small number of large shells (Pl. 1, fig. 14). Concentric growth lines well developed, becoming stronger toward anterior margin to develop into coarse, concentric lamellae in some larger shells (Pl. 1, figs. 18, 21). Minute perforations present along costal crests, especially near anterior margin (Pl. 1, figs. 7, 14; Pl. 3, figs. 1, 2, 4), probably as traces of non-preserved small, hollow spines or pustules.

Fig. 6. Shell dimensions of *Dinorthis occidentalis* (Okulitch, 1943), sample from GSC loc. O-37130, Gunn Member, Stony Mountain Formation. Note that shells are consistently wider than long throughout ontogeny.

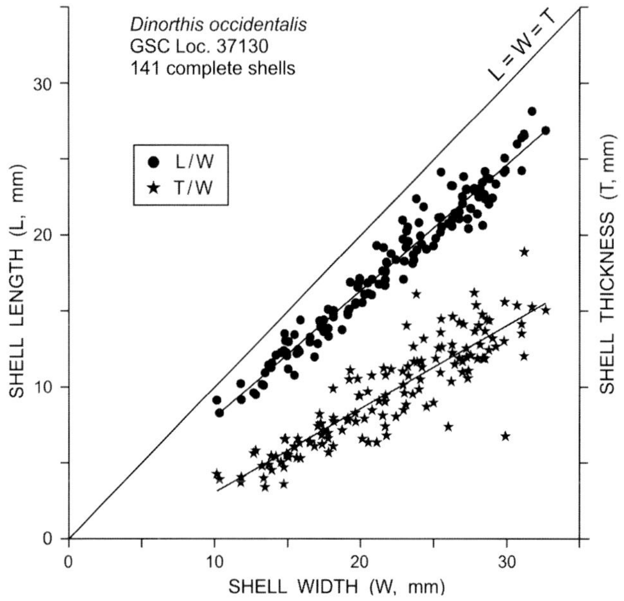

Fig. 7. Shell dimensions of *Dinorthis occidentalis* (Okulitch, 1943), sample from GSC loc. O-37130, Gunn Member, Stony Mountain Formation. Note consistent ratio of hinge width / shell width (W_1/W) throughout ontogeny.

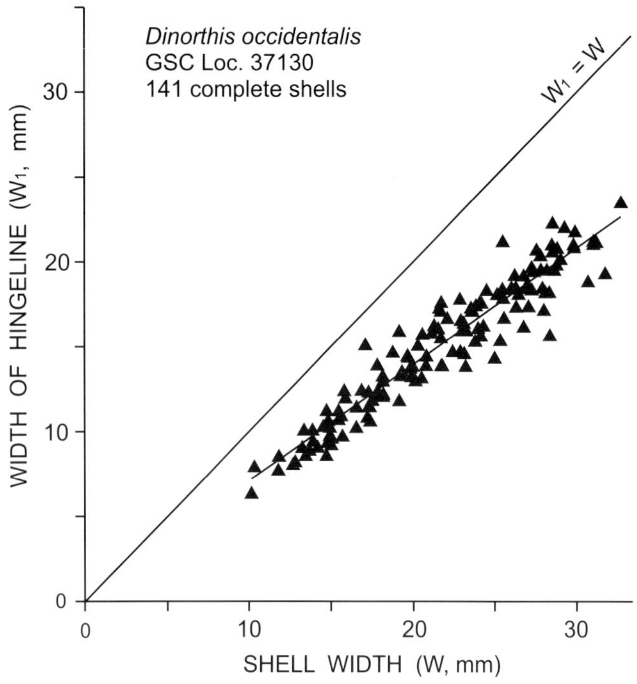

Statistics of shell measurements (mm):

GSC loc. O-37130	L	W	L/W	T	W_1	W_1/W
AVG	17.84	21.85	0.82	9.56	15.23	0.70
STD	4.69	5.52	0.05	3.42	3.97	0.06
MIN	8.27	10.15	0.70	3.40	6.40	0.60
MAX	28.16	32.69	0.95	18.89	23.56	0.89

Ventral interior. Teeth strong, massive; dental plates thick, subparallel, extending anteriorly as low, postero-lateral bounding ridges of muscle field; secondary shell deposit filling lateral cavities, thus fusing dental plates to lateral shell wall (Figs. 8, 9). Muscle field subrectangular to subtrapezoidal in outline (Pl. 1, figs. 15, 17; Pl. 2, figs. 1–3, 7, 11, 12), occupying 36–40% of shell width and 45–57% of shell length; adductor muscle scars elongate oval or cordate, deeply impressed and well defined, with variously striated floor (Pl. 2, figs. 5, 12), located in posterior-medial part of muscle field, immediately anterior of weak delthyrial platform. Diductor scars bilobate, usually wider anteriorly, enclosing adductor scars laterally and anteriorly, with fine, longitudinal trans-muscle ridges and striae. Marginal crenulations strong, evenly distributed along entire lateral and anterior shell margins.

Dorsal interior. Notothyrial platform well developed; cardinal process high, strong, ridge-like, with well-differentiated, swollen, trifid myophore (Pl. 2, figs. 9, 10; Fig. 8, 1.2 mm from apex), occupying almost entire notothyrium and projecting into delthyrium of opposite valve; basal portion of cardinal process isolated completely from brachiophore bases. Brachiophores robust, diverging from each other at about 80–90°, forming part of lateral bounding ridges of notothyrial platform, becoming free as short, stubby processes anterior of highly elevated notothyrial platform (Pl. 2, figs. 8–10). Adductor muscle field marked by pair of subrectangular scars divided by thick, strong median ridge; floor of muscle field bearing fine, reticulate ridges (Pl. 2, figs. 8, 10; see also Jin et al. 1997). Marginal crenulations well developed.

Remarks. Examination of several hundred specimens (sample GSC loc. O-37130) of *Dinorthis occidentalis* shows a wide range of infraspecific variations:

1. Most of the smaller shells are biconvex or even slightly ventribiconvex. With increasing shell size, the dorsal valve becomes more strongly convex than the ventral valve, whereas the antero-medial part of the ventral valve becomes more strongly depressed into a broad, shallow, sulcus-like structure.
2. The dorsal interarea is usually orthocline but becomes apsacline in very strongly convex valves.
3. The hinge teeth are generally delicate and platy in relatively small shells, becoming more robust with ontogeny. The dental plates, however, remain thick and strong throughout ontogeny.
4. The ventral muscle field shows a wide range of variation in outline, particularly in its anterior portion (compare Pl. 2, figs. 1, 5). The diductor scars are prominently bilobate, forming a prominent medial notch (emargination) at the anterior margin of the muscle filed. The depth of the antero–medial notch varies greatly, ranging from broad and shallow to those with a narrow medial gap (Pl. 2, figs. 1, 12). The adductor scars, however, are rarely enclosed completely by the diductor scars at the notch.

Fig. 8. Transverse serial sections of *Dinorthis occidentalis* (Okulitch, 1943), hypotype GSC 117793 (mature shell: *L* = 20.9 mm, *W* = 27.8 mm, *T* = 11.8 mm), from GSC loc. O-37130, Gunn Member, Stony Mountain Formation.

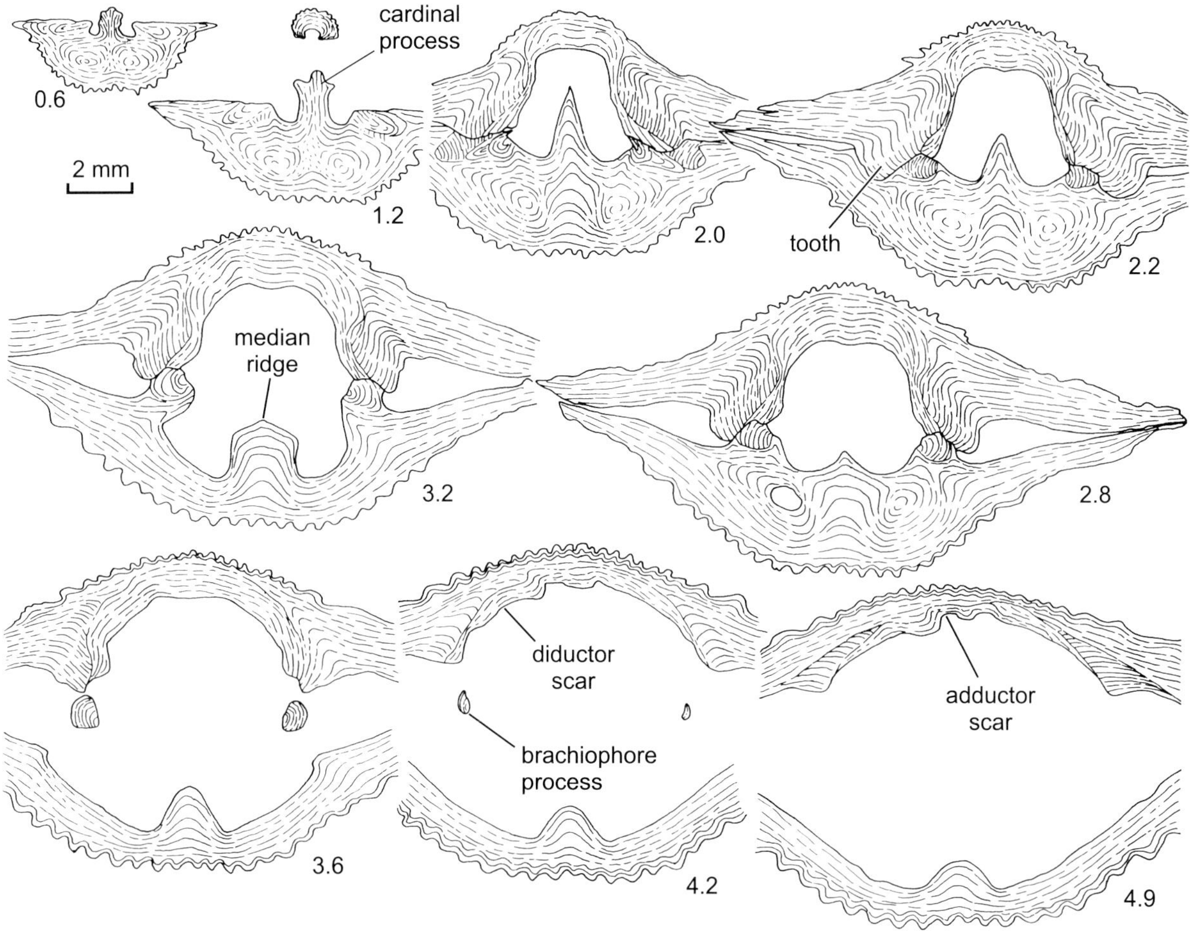

Fig. 9. Transverse serial sections of *Dinorthis occidentalis* (Okulitch, 1943), hypotype, GSC 117800 (juvenile shell: *L* = 12.8 mm, *W* = 16.4 mm, and *T* = 6.6 mm), GSC loc. O-37130, Gunn Member, Stony Mountain Formation.

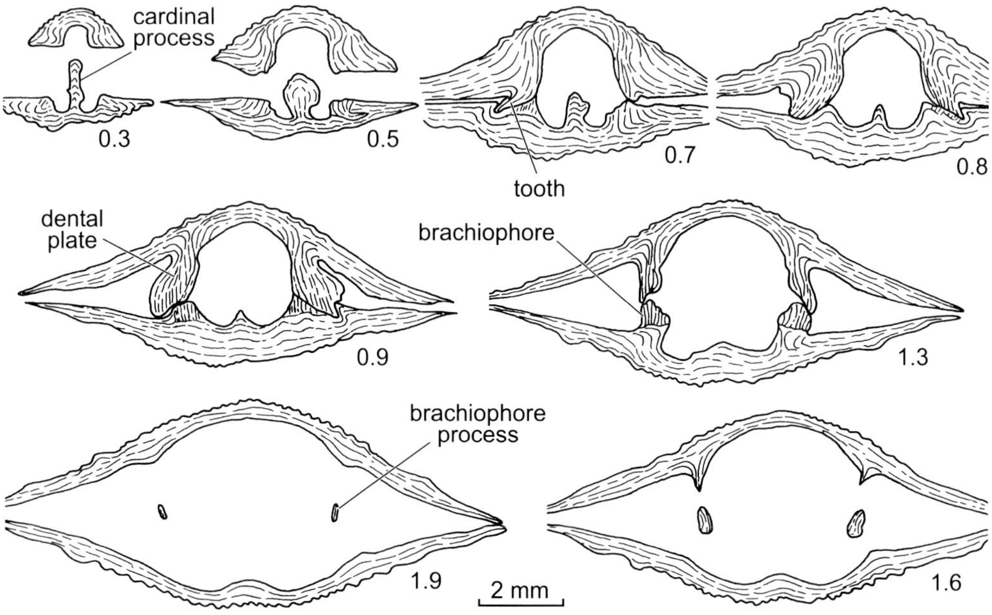

Dinorthis occidentalis is one of the most abundant and easily recognized elements of the Red River Brachiopod Fauna (Okulitch 1943; Ross 1957; Macomber 1970; Jin et al. 1997). Because of its general abundance and wide distribution in North America, this species has been studied by many authors, with minor discrepancies with regards to its generic or subgeneric assignment (see synonymy list). Because of its strong dorsibiconvexity, particularly at the adult stage, the species does not belong to *Pionorthis*, which is characterized by a nearly equibiconvex shell.

In his study of the brachiopods from the uppermost part of the Bighorn Dolomite in the Rock Creek area of Wyoming, Ross (1957) recognized five species: *Dinorthis* (*Plaesiomys*?) cf. *D.* (*P.*) *occidentalis* Ladd, *Dinorthis* (*Pionorthis*?) cf. *D.* (*P.*) *occidentalis Okulitch*, *Dinorthis* (*Pionorthis*?) sp., *Dinorthis*? sp., and *Dinorthis*? (*Pionorthis*?) sp., mostly on the basis of detailed external differences, such as shell convexity, length of hingeline, comparative density of costae, and the presence or absence of rare costal bifurcations. Examination of the southern Manitoba material indicates that the morphological differences noted by Ross are infraspecific variations, and therefore, all the five species described by Ross should be assigned to *Dinorthis occidentalis*.

Family PLECTORTHIDAE Schuchert and LeVene, 1929

Genus *Gnamptorhynchos* Jin, 1989

Type species. *Platystrophia regularis* var. *globata* Twenhofel, 1928, Prinsta Member, Ellis Bay Formation, latest Ordovician (Hirnantian), Anticosti Island. As proposed by Jin and Zhan (2000), the originally designated type species, *Gnamptorhynchos inversum* Jin, 1989, should be replaced by its senior synonym, *Platystrophia regularis* var. *globatum*.

Gnamptorhynchos manitobensis Jin and Zhan, 2000

Pl. 3, figs. 6–20; Figs. 10–12

2000 *Gnamptorhynchos manitobensis* Jin and Zhan, p. 989, figs. 4.1–4.15.

Types. Five specimens of original designation: holotype, GSC 117741 (Pl. 3, figs. 16–20); paratypes, GSC 117739 (immature shell), GSC 117740, GSC 117742 (serially sectioned), and GSC 117743 (serially sectioned). All from the Selkirk Member, Red River Formation (MMMN I-2164), southern Manitoba.

Other material examined. Selkirk Member, Gillis Quarry (I-2164): 12 conjoined shells, 3 dorsal and 1 ventral valves.

Description. Shell medium-sized to large, transversely subquadrate to subelliptical (Fig. 10), with average length 16.0 mm (maximum 20.5 mm), width 19.6 mm (maximum 26.0 mm), and thickness 13.8 mm (maximum 20.3 mm). Lateral profile biconvex to slightly dorsibiconvex. Dorsal fold and ventral sulcus well developed, both originating from the umbo, becoming wider anteriorly to occupy slightly more than two-fifths of shell width. Hingeline straight, attaining about three-fourths of shell width (Fig. 11). Cardinal extremities rounded. Greatest shell width reached near mid-length. Both ventral and dorsal umbones strongly arched, with curved beaks. Ventral interarea sharply defined, apsacline, with maximum height of 2 mm; delthyrium open. Dorsal interarea low (generally not exceeding 1 mm), orthocline. Costae simple, evenly spaced, with subangular to subrounded crests. Each shell flank bearing 10–12 costae in adult forms but 7–8 in immature ones. Number of costae on fold and in sulcus variable (independent of shell size), ranging from 4–6 and 3–5, respectively. Concentric growth lines well developed over entire shell surface; imbricating growth lamellae present only near anterior margin of relatively large shells.

Shell measurements (*mm*):

	L	*W*	*T*	W_1	W_2	*L*/*W*	W_1/W	W_2/W
GSC 117739	10.2	12.2	8.3	8.9	5.1	0.83	0.73	0.42
GSC 117740	15.4	19.2	12.7	15.2	8.3	0.8	0.79	0.44
GSC 117741	20.5	26.1	20.3	19.5	12.4	0.79	0.75	0.48
GSC 117742	19.4	22.8	17.4	17.7	10.8	0.85	0.78	0.47
GSC loc. C-205935	14.3	18.6	11.3	12.6	7.8	0.77	0.68	0.42
GSC loc. C-205935	16.3	18.6	12.5	14.1	8.1	0.87	0.75	0.44

Fig. 10. Shell dimensions of *Gnamptorhynchos manitobensis* Jin and Zhan, 2000, sample from Garson Quarry (GSC loc. C-205935), Selkirk Member, Red River Formation. Note strong globosity of the shells (*T*/*W* ratios).

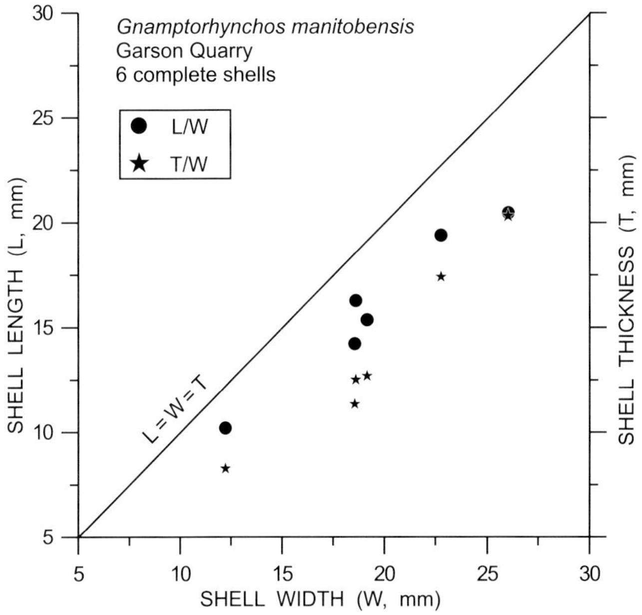

Interior. Teeth small; dental plates thick, subparallel, buried apically into secondary shell thickening to fuse with lateral shell walls, extending anteriorly to become lateral bounding ridges of muscle field (Fig. 12). Ventral muscle field elongate, subquadrate, well impressed in its posterior portion, becoming slightly elevated anteriorly and confined by high bounding ridges. Cardinal process low, thin, ridge-like, confined to posterior part of notothyrial cavity. Secondary shell deposits well developed in umbonal area. Median ridge short, formed by secondary shell thickening, dorsally supporting small, septalium-like notothyrial platform. Brachiophore bases massive, extending anteriorly into platy brachiophore processes. Adductor scars sharply impressed, becoming wider anteriorly, bearing low median ridge, and confined by conspicuous lateral bounding ridges.

Fig. 11. Comparison of hinge width / shell width (W_1/W) ratios of *Gnamptorhynchos manitobensis* Jin and Zhan, 2000, *G. globatum* (Twenhofel, 1928), and *Platystrophia regularis* Shaler, 1865. Note that the average W_1/W ratio of *G. manitobensis* is intermediate between those of *G. globatum* and *P. regularis*.

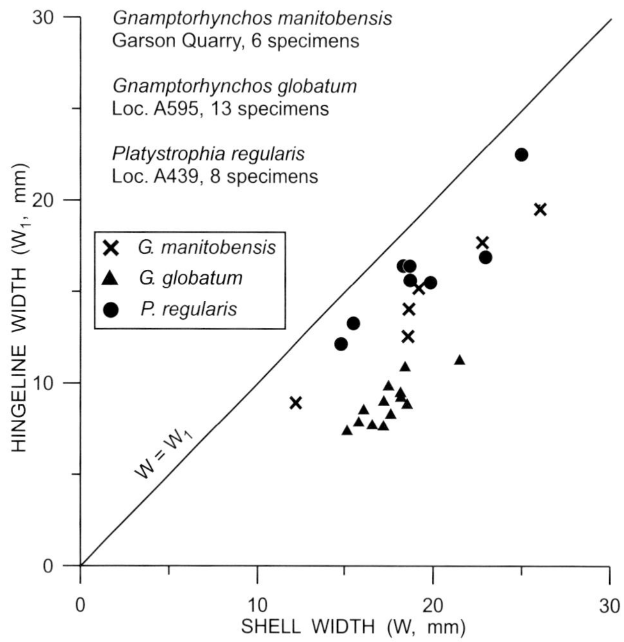

Discussion. *Gnamptorhynchos manitobensis* differs from the other known congeneric species, *G. globatum* and *G. selliseptalicium*, in having a wider hingeline and a greater number of costae on the fold and in the sulcus. In the studied collection, 6 of the 12 well-preserved specimens have six costae on the fold (five in the sulcus), 3 shells have four on the fold (three in the sulcus), 2 have five on the fold (four in the sulcus), and 1 ventral valve has seven costae in the sulcus. *Gnamptorhynchos globatum* and *G. selliseptalicium* usually have four (rarely three or five) costae on the fold and three (rarely two or four) in the sulcus.

Fig. 12. Transverse serial sections of *Gnamptorhynchos manitobensis* Jin and Zhan, 2000, paratype, GSC 117743 (mature shell: L = 18.7 mm, W = 22.6 mm, T = 17.6 mm), Garson Quarry (GSC loc. C-205935), Selkirk Member, Red River Formation.

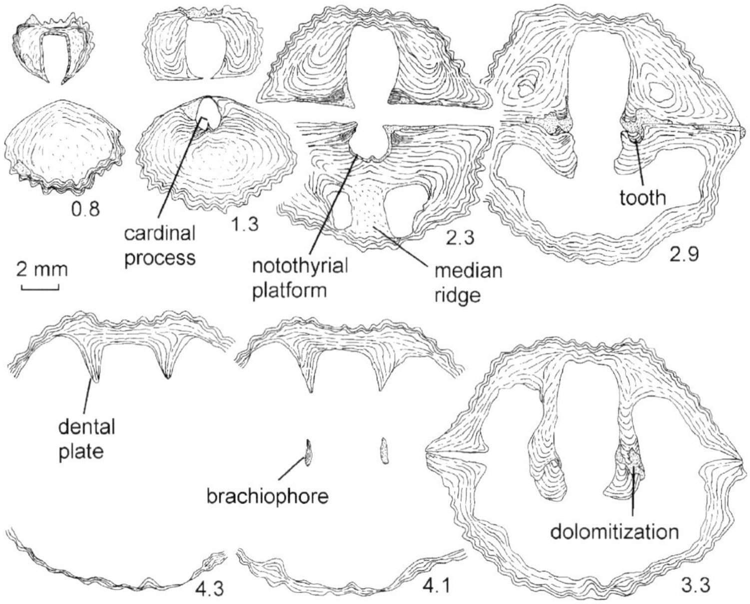

Internally, *Gnamptorhynchos manitobensis* differs from the type species, *G. globatum*, in lacking accessory teeth and sockets and in having a unilobate cardinal process. The cardinal process of the new species is incipient, much like the cardinal process of the type species of *Platystrophia biforatus* (von Schlotheim, 1820). The Manitoba species is assigned to *Gnamptorhynchos* rather than to *Platystrophia* on account of its strongly globular shell at adult stage and septalium-like notothyrial platform. The W_1/W ratio of *Gnamptorhynchos manitobensis* is intermediate between that of *G. globatum* and that of *Platystrophia regularis* (Fig. 11).

Superfamily DALMANELLOIDEA Schuchert, 1913

Family PLATYORTHIDAE Harper, Boucot, and Walmsley, 1969

Genus *Diceromyonia* Wang, 1949

Type species. *Orthis tersus* Sardeson, 1892, pp. 331–332, pl. 5, figs. 11–13; "The specimens figured were collected from the Cincinnati group, at Wilmington, Illinois" (Sardeson 1892, p. 332). The specimens are most likely from the upper Maquoketa Formation, Richmondian (Amsden 1974; Harper et al. 1969).

Remarks. According to Wang (1949), *Diceromyonia* can be distinguished from all other Ordovician dalmanellids by its long ventral diductor scars that completely enclose the adductor scars. Harper et al. (1969) proposed a new subfamily, the Platyorthinae, within the Family Rhipidomellidae and assigned *Diceromyonia* to their new subfamily. This was followed by Amsden (1974). Howe (1965, 1988), however, prefer to retain the genus in the Family Dalmanellidae. In the revised brachiopod volumes of the Treatise on Invertebrate Paleontology, Williams and Harper (2000) raised the Platyorthinae to familial status and included *Diceromyonia* in the family. This is followed here, except to note that some forms of *Diceromyonia* may appear to be similar to certain species of *Dalmanella* or *Onniella*. For example, *Dalmanella testudinaria* (Dalman, 1828) shows a high degree of similarity to *Diceromyonia storeya*, particularly in the shell shape, costae, and cardinalia, but *Diceromyonia storeya* differs in its more convex dorsal valve, much larger and longer ventral muscle field, and more poorly developed fulcral plates. *Onniella meeki* (Miller, 1875) from the Arnheim Formation of Tennessee (Howe 1988) also shows superficial similarity to *Diceromyonia storeya*, but differs in its more strongly divergent brachiophores, subcircular dorsal muscle field, and the presence of ventral muscle bounding ridges.

Diceromyonia storeya (Okulitch, 1943)

Pl. 4, figs. 1–20; Pl. 5, figs. 1–5; Figs. 13–15

1943 *Dalmanella storeya* Okulitch, p. 70, pl. 1, figs. 1–4.

1957 *Diceromyonia storeya* (Okulitch); Ross, p. 487, pl. 41, figs. 5, 6, 9, 12, 16.

1960 *Diceromyonia* cf. *storeya* (Okulitch); Brindle, pl. 4, fig. 6.

1970 *Diceromyonia storeya* (Okulitch); Macomber, p. 436, pl. 77, figs. 1–43.

1997 *Diceromyonia storeya* (Okulitch); Jin et al., p. 22, pl. 3, figs. 1–22; pl. 4, figs. 1–3.

Types. The holotype, GSC 1362 (Pl. 5, figs. 1–5), and paratypes (GSC 1362a, b) are from the Gunn Member, Stony Mountain Formation (Richmondian), Stony Mountain, Manitoba.

Additional material examined. Penitentiary Member, (GSC loc. O-27186, C-205933, and 55-60; UM EPD-1995): 18 conjoined shells, 7 dorsal and 5 ventral valves; 39 dorsal internal and 29 external, 42 ventral internal and 28 external moulds, and 6 complete internal moulds.

Gunn Member (GSC loc. O-27185, O-37130, C-205926, C-205928, C-205930, and C-205931): 528 conjoined shells, 233 dorsal and 240 ventral valves.

Selkirk Member (I-900, I-2165): 6 conjoined shells, 1 dorsal and 2 ventral valves, and 1 dorsal external mould.

Dog Head Member (I-2418, I-2653): 1 conjoined shell and 1 ventral internal mould.

Description. Shell medium-sized, transversely subelliptical (average length/width ratio 0.88), with average length of 11.0 mm (maximum 15.0 mm), width 12.6 mm (maximum 17.5 mm), and thickness 5.9 mm (maximum 9.3 mm; Fig. 13). Lateral profile nearly equibiconvex to slightly ventribiconvex, with greatest thickness and convexity at about one-third of shell length from apex. Hingeline relatively short for dalmanellids, attaining less than half of maximum shell width, with W_1/W ratio decreasing notably with ontogeny (Fig. 14). Cardinal extremities rounded. Ventral umbo carinate, with carina becoming gradually flattened toward anterior margin. Dorsal umbo strongly sulcate, gradually developing into somewhat broader and shallower sulcus anteriorly. Ventral interarea apsacline, with minute, slightly incurved beak; delthyrium open; dorsal interarea much lower than ventral interarea, invariably orthocline with open notothyrium. Shell costae strong, averaging three costae per 1 mm at 5 mm from apex, increasing anteriorly by bifurcation (common) and intercalation (rare).

Fig. 13. Shell dimensions of *Diceromyonia storeya* (Okulitch, 1943), sample from GSC loc. O-37130, Gunn Member, Stony Mountain Formation. Note wide range of variation in shell convexity (*T*/*W*), especially in relatively large individuals.

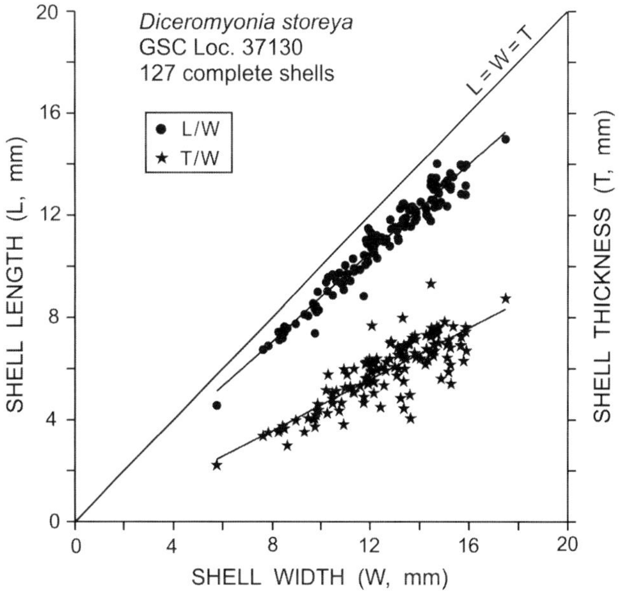

Fig. 14. Shell dimensions of *Diceromyonia storeya* (Okulitch, 1943), sample from GSC loc. O-37130, Gunn Member, Stony Mountain Formation. Note that the W_1/W ratio decreases notably with ontogeny.

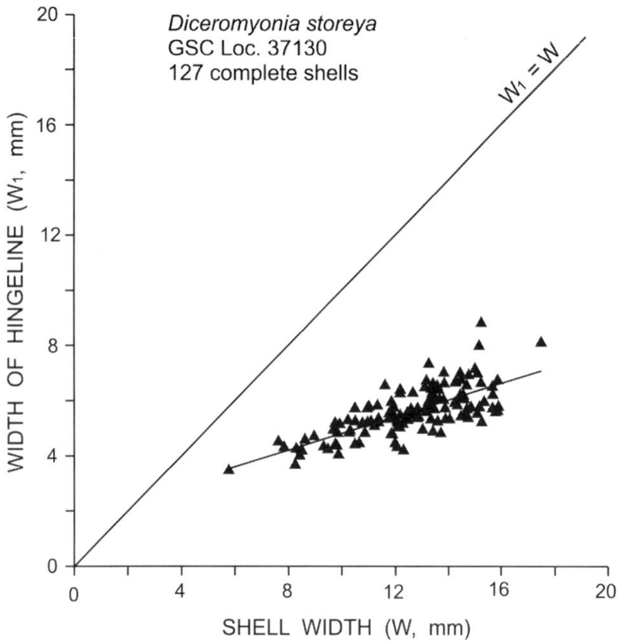

Primary costae starting from apices of both valves, usually bifurcating twice or three times, with first-order bifurcation occurring at about mid-length and second-order bifurcation at anterior fourth or fifth of shell; costal crests at bifurcating points usually unequal, with thinner-crested branch usually on medial side of thicker-crested branch in dorsal valve, and vice versa in ventral valve (Pl. 4, figs. 1, 2, 11, 13, 14, 18; Pl. 5, figs. 1, 2). Dorsal valve invariably bearing prominent median costa (Pl. 4, figs. 1, 11, 13; Pl. 5, fig. 1). Concentric growth lines very fine, hardly visible to naked eye; several strong growth lamellae present around shell margin of relatively large shells, with additional set occurring also near mid-length in some shells (Pl. 4, figs. 13, 14). One row of minute perforations present on each costal crest, especially well developed toward shell margin (Pl. 4, figs. 9, 13, 14, 16, 18), probably as traces of non-preserved hollow spines or pustules.

Statistics of shell measurements (mm)

GSC loc O-37130	*L*	*W*	*T*	W_1	*L/W*	W_1/W
AVG	11.00	12.56	5.85	5.58	0.876	0.450
STD	1.94	2.18	1.31	0.90	0.035	0.060
MIN	4.56	5.77	2.21	3.46	0.751	0.341
MAX	14.99	17.48	9.32	8.81	0.963	0.600

Microscopic shell structure. Punctae extremely fine, hardly visible at scale of Fig. 15.

Ventral interior. Teeth strong, massive, with prominent hinge fossettes near their bases on inner sides; dental plates short, extending anteriorly to become postero-lateral bounding ridges of muscle field. Muscle field deeply impressed, usually extending for more than half of valve length. Diductor muscle scars elongate oval, scalloped (Pl. 4, figs. 19, 20), forming small, narrow medial notch (emargination) at anterior margin of field; narrow scalloped lobe on each postero-lateral side of muscle field probably representing adjustor muscle scars. Adductor scars longitudinally elliptical, located in postero-medial part of muscle field, immediately anterior of deep delthyrial cavity, and completely enclosed by diductor scars. Low median ridge (myophragm) of muscle field usually double-crested, originating from anterior part of adductor scars, becoming slightly higher anteriorly, and ending at anterior end of muscle field. Inner shell margin marked by fine crenulations, slightly longer and stronger in medial part of shell margin (Pl. 4, figs. 6, 10, 19).

Dorsal interior. Cardinalia robust but relatively small in size, occupying about 22% of shell length and 28% of shell width; cardinal process high, located at posterior end of slightly elevated notothyrial platform, and projecting ventrally into delthyrial cavity; myophore of cardinal process usually trifid, with median lobe being much higher than lateral lobes. Brachiophores strong, diverging from each other at about 70–80°, completely separated from cardinal process by sharp grooves, and acting as inner bounding ridges of deep sockets. Fulcral plates usually present but weak (Pl. 4, figs. 6, 10, 12). Adductor muscle field shallow but well delimited, lacking bounding ridges; two pairs of adductor scars divided by thick median ridge (myophragm), originating immediately anterior of notothyrial platform; anterior pair of adductor scars subcircular, extending near mid-length of valve, and posterior pair trapezoidal in shape, smaller and narrower than anterior pair. Crenulations on inner shell margin corresponding to those of ventral valve.

Remarks. Shells of *Diceromyonia storeya* from the upper Bighorn Dolomite (Ross 1957) and from the Churchill River Group of the Hudson Bay Lowlands (Jin et al. 1997) are identical to those from southern Manitoba in their shape, costae, long and scalloped

Fig. 15. Serial sections of *Diceromyonia storeya* (Okulitch, 1943), hypotype, GSC 117802 (mature shell: L = 12.8 mm, W = 14.3 mm, T = 7.0 mm), GSC loc. O-37130, Gunn Member, Stony Mountain Formation.

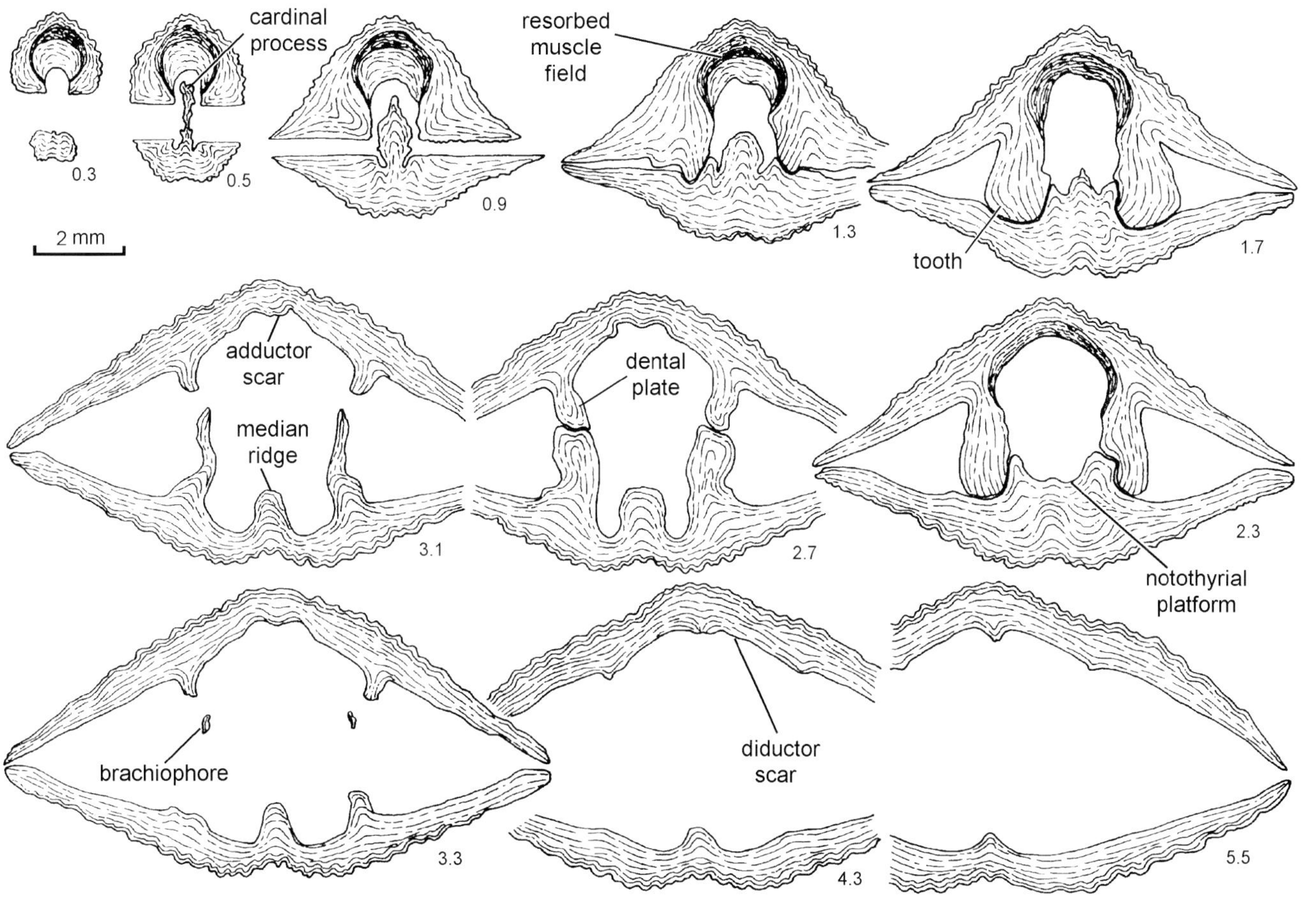

ventral muscle field, robust teeth, and cardinalia. *Diceromyonia* cf. *D. ignota* (Ross 1957) from the wer Stony Mountain Formation of eastern Montana and *Diceromyonia* aff. *D. terxsa* (Ross 1957) from the uppermost Bighorn Dolomite of Wyoming are nearly identical to *D. storeya* from southern Manitoba, except for their slightly longer ventral muscle field in some specimens.

Order STROPHOMENIDA Öpik, 1934

Superfamily PLECTAMBONITOIDEA Jones, 1928

Family SOWERBYELLIDAE Öpik, 1930

Genus *Thaerodonta* Wang, 1949

Type species. *Thaerodonta aspera* Wang, 1949. Upper Elgin Member (Maysvillian), Maquoketa Formation, Iowa.

Remarks. Cocks and Rong (1989) listed Thaerodonta as a junior synonym of Sowerbyella (Eochonetes) Reed, 1917, on the basis of both genera having hinge denticles in the dorsal valve and corresponding hinge fossettes in the ventral valve. Previously, Eochonetes was distinguished from other sowerbyellids in its ventral hingeline being perforated by oblique canals. Cocks and Rong regarded these canals as insignificant taxonomic characters at the generic or subgeneric levels because they have been found only in about half the Eochonetes populations. The presence of hinge denticulation (dorsal denticles and ventral fossettes) was regarded by Howe (1972) as a diagnostic feature of Thaerodonta, although, as will be discussed below, denticulations do not occur in all members of the genus, contrary to Howe's (1972, p. 443) statement that "regardless of age, shape, size, variation, or species, . . . each pedicle valve of Thaerodonta bears fossettes and each brachial valve displays the corresponding denticles." Other internal features commonly used as diagnostic characteristics in the sowerbyellids, such as the side septa of the dorsal valve, also show considerable variations (Howe 1965; Macomber 1970; Mitchell 1977; Cocks and Rong 1989).

Despite wide range of variations in a number of diagnostic characters, *Thaerodonta* can be distinguished easily from *Sowerbyella* and *Eochonetes* when large populations are examined. Following Potter and Boucot (1992) and Jin et al. (1997), the genus *Thaerodonta* is retained as an independent genus herein.

Thaerodonta clarksvillensis (Foerste, 1912)

Pl. 4, figs. 21, 22; Pl. 5, figs. 6–8

1912 *Plectambonites rugosa-clarksvillensis* Foerste, p. 127, pl. 1, figs. 7a–c; pl. 10, figs. 7a–d.

1944 *Sowerbyella clarksvillensis* (Foerste); Cooper, p. 225, pl. 128, figs. 39, 40.

1961 *Sowerbyella rugosus clarksvillensis* (Foerste); Caster et al., pl. 7, figs. 25, 26.

1970 *Thaerodonta* aff. *T. clarksvillensis* (Foerste); Macomber, p. 439, pl. 78, figs. 1–15.

1979 *Thaerodonta clarksvillensis* (Foerste); Howe, p. C4, pl. 2, figs. 14–17.

1988 *Thaerodonta clarksvillensis* (Foerste); Howe, p. 206, figs. 2.8, 2.13.

1988 *Thaerodonta recedens* (Sardeson); Howe, p. 214, figs. 2.9–2.12; 2.14–2.17, 10, 11.

1997 *Thaerodonta clarksvillensis* (Foerste); Jin et al., p. 23, pl. 4, figs. 10–16.

Types. Under discussion of *Plectambonites rugosa* (Meek, 1873), Foerste (1912, p. 123) proposed a subspecies, *Plectambonites rugosa-clarksvillensis*, "to designate the large specimens from the lower third of the Clarksville division of the Waynesville bed, in Ohio and Indiana." Among the specimens he figured, only the dorsal valve (Foerste 1912, pl. 10, fig. 7a) was cited as from the Clarksville bed (Richmondian), Ohio, and should be regarded as the lectotype.

Southern Manitoba material. Penitentiary Member (GSC loc. 55-60): 9 dorsal internal and 1 external, and 2 ventral internal moulds.

Selkirk Member (I-2472, I-2766): 2 conjoined shells.

Cat Head Member (UM No. S168): 8 conjoined shells.

Dog Head Member (I-2415, I-2419): 55 conjoined shells.

Description. Shell small, transversely semielliptical, 4.2–5.5 mm long and 8.0–10.7 mm wide with average length/width ratio about 0.53. Lateral profile concavo-convex, deepest in central part of shell. Hingeline straight, being widest part of shell; cardinal extremities subrectangular or slightly pointed. Both ventral and dorsal interareas low (<0.5 mm in height), apsacline and anacline respectively; delthyrium and notothyrium open. Entire shell covered by sharply differentiated parvicostellae, averaging 5–7 fine costellae (papillae) between two adjacent coarser costellae.

Microscopic shell structure. Large pseudopunctae, extending into irregular tubercles (pustules) around marginal and postero-lateral areas of shell (Pl. 4, fig. 21; Pl. 5, figs. 6, 7).

Ventral interior. Teeth small but relatively robust; dental plates low, extending anteriorly to become lateral bounding ridges of muscle field. Five or six small hinge fossettes present in medial portion (between apex and cardinal extremity) on each side of hingeline. Muscle field bilobate, well defined, occupying 35% of shell width and 64% of shell length; pair of diductor muscle scars diverging from each other at about 45°, with relatively large, wide, medial notch occupying approximately half of muscle-field length; adductor muscle scars small, oval-shaped, slightly deeper than diductor scars, extending posteriorly into round cavity underneath delthyrial platform, divided by median septum. Median septum beginning immediately anterior of delthyrial platform, becoming stronger abruptly at anterior end of adductor scars, attaining greatest height at its anterior end, and then bifurcating into pair of low ridges as antero-medial bounding ridges of diductor scars (Pl. 4, fig. 21; Pl. 5, fig. 6). Vascular markings saccate, with vascula media originating from anterior ends of diductor scars.

Dorsal interior. Cardinal process undercut, bearing trifid myophore with median lobe being highest, laterally continuous with socket ridges (Pl. 5, fig. 7). Socket ridges straight, short, thick, and divergent from each other antero-laterally at about 120°. Five or six denticles (accessory teeth) present in medial portion on each side of hingeline, corresponding to ventral dental fossettes. Diductor muscle field suboval to subcircular in outline, attaining about 40% of shell width and 65% of shell length; pair of central side septa relatively long, high, and thin, originating immediately anterior of socket ridges, diverging from each other anteriorly at about 15–20°, extending anteriorly for slightly over three-fourths of valve length, and attaining greatest height near anterior margin of muscle field; pair of lateral side septa (transmuscle ridges) variably developed but generally weak, separating each adductor muscle scar into inner and outer parts.

Remarks. Howe (1988) established *Thaerodonta recedens* on the basis of material from the Arnheim Formation of Tennessee and distinguished it from *Thaerodonta clarksvillensis* by its narrower brachiophores. However, Howe (1988) also noted gradational change in the divergence angle of brachiophores between *T. clarksvillensis* of Ohio and *T. recedens* from Tennessee and regarded the two as conspecific. Observation of Howe's figured specimens and the specimens used in this study confirms the existence of a wide range of variation in the brachiophore divergence angle (for example, compare Howe's fig. 2.12 with fig. 2.17). All the illustrated specimens of *Thaerodonta* aff. *T. clarksvillensis* from the Rock Creek Beds and the Hunt Mountain Beds in the Bighorn Mountains of northern Wyoming (Macomber 1970) appear identical to Howe's specimens of *T. recedens*. Both of these species are regarded as conspecific with *T. clarksvillensis*.

Superfamily STROPHOMENOIDEA King, 1846

Family STROPHOMENIDAE King, 1846

Genus *Strophomena* Rafinesque *in* de Blainville, 1824

Type species. *Leptaena planumbona* Hall, 1847. Trenton Limestone, Hudson River Group (Caradoc), Cincinnati, Ohio.

Remarks. Cocks (1990) proposed *Leptaena planumbona* as the type species to replace the poorly known *Strophomena rugosa* de Blainville, 1824, which is probably conspecific with *L. planumbona*. This proposal was formalized by ICZN Opinion 1671 (1992). Despite the detailed study of Rong and Cocks (1994), confusion remains about the true identity of *Strophomena planumbona*, especially with regards to its dorsal internal structures and its type locality and type stratum. Hall's (1847) broad statement of the type stratum as "a position equivalent to that of the Trenton Limestone . . . in Cincinnati and Oxford (Ohio), Madison (Indiana), and Marysville (Kentucky)" was taken at face value by Rong and Cocks (1994), who assigned a Caradoc age to the type species. In the late nineteenth century, Trenton Limestone was used for a much wider range of Ordovician strata than in modern usage, and could include rocks of late Caradoc to mid Ashgill age. In the American midcontinent, *Strophomena planumbona* was largely confined to Maysvillian and Richmondian strata. In the Cincinnati type area, in particular, *Strophomena planumbona* occurs only in Richmondian strata of the Arnheim, Waynesville, and Liberty formations (Davis 1985). In other areas of North America, the species is known from the following strata:

1. Brainard Member (Richmondian) of the Maquoketa Formation of Iowa (Wang 1949);

2. Viola Formation (lower Maysvillian) of Oklahoma (Alberstadt 1973);
3. Arnheim Formation of Tennessee (Howe 1988);
4. Surprise Creek and Caution Creek formations (late Maysvillian – early Richmondian), Hudson Bay Lowlands (Jin et al. 1997);
5. Vauréal Formation (Richmondian), Anticosti Island, Québec (Dewing 1999).

Available information indicates that *Strophomena planumbona* is largely of Ashgill age and most likely confined to Richmondian strata in the Cincinnati type area.

As pointed out by Rong and Cocks (1994) and Dewing (1999), there has been a great deal of confusion over the dorsal internal structures of *Strophomena planumbona* with those of *S. vetusta* (James, 1874) and *S. filitexta* (Hall, 1847), mainly because Hall (1847) did not illustrate any dorsal interiors in his initial or subsequent descriptions of the species. Muir-Wood and Williams (1965) illustrated a dorsal interior of *S. planumbona* (from Ohio) with four long, strong, and straight transmuscle septa, extending for over two-thirds of the valve length; these septa are similar to those of *S. vetusta*. This seems to agree with Pope's (1976, p. 176) definition of the *Strophomena*-type transmuscle septa. The two dorsal valves illustrated by Rong and Cocks (1994), however, are drastically different, showing short and unusually weak transmuscle septa confined to the posterior half of the valve; the two central side septa are arched laterally, enclosing a suboval area, as is characteristic of *Holtedahlina*, *Pentlandina*, and *Katastrophomena* (Cocks 1968; Pope 1976). Serial sections of *S. planumbona* from Ohio and Anticosti Island confirm the presence of four high, sharp, closely spaced transmuscle septa (Dewing 1999). The dorsal interior of true *Strophomena*, therefore, should have two pairs of long, strong, relatively straight transmuscle septa, as illustrated by Muir-Wood and Williams (1965) and typified in *S. vetusta*. The two dorsal valves illustrated by Rong and Cocks (1994) as *S. planumbona* require further study to determine their identity.

Strophomena planumbona (Hall, 1847)

Pl. 6, figs. 3–8; Pl. 7, figs. 1–4

1847 *Leptaena planumbona* Hall, p. 112, pl. 31B, figs. 4a–e.

1859 *Strophomena planumbona* (Hall); Hall, p. 54, fig. 7.

1873 *Strophomena planumbona* (Hall); Meek, p. 79, pl. 6, figs. 3a–h.

1892 *Strophomena planumbona* (Hall); Hall and Clarke, pl. 9, figs. 15–17; pl. 9A, figs. 8, 9.

1912 *Strophomena planumbona* (Hall); Foerste, p. 73, pl. 4, fig. 3; pl. 8, figs. 1a–e; pl. 9, figs. 3a, b.

1924 *Strophomena planumbona* (Hall); Foerste, p. 117, pl. 12, figs. 1a–c.

1944 *Strophomena planumbona* (Hall); Cooper, p. 337, pl. 130, figs. 21–24.

1949 *Strophomena planumbona* (Hall); Wang, p. 23, pl. 6D, figs. 1–7.

1961 *Strophomena planumbona* (Hall); Caster et al., pl. 8, figs. 8–11.

1973 *Strophomena planumbona* (Hall); Alberstadt, p. 45, pl. 5, figs. 13, 14.

1985 *Strophomena planumbona* (Hall); Davis, pl. 8, figs. 8–11.

1988 *Strophomena planumbona* (Hall); Howe, p. 216, figs. 6.30, 6.31.

1994 *Strophomena planumbona* (Hall); Rong and Cocks, p. 677, pl. 1, figs. 1–6, 7?, 9?, 10, 11.

1997 *Strophomena planumbona* (Hall); Jin et al., p. 24, pl. 9, figs. 9–12; pl. 10, figs. 1, 2.

1999 *Strophomena planumbona* (Hall); Dewing, p. 24, pl. 8, figs. 1–13.

Type specimen. Lectotype, AMNH 30247 (American Museum of Natural History), selected by Rong and

Cocks (1994) from Hall's original type lot. Hall's figured specimens came from Cincinnatian rocks of Ohio, with the precise type stratum unknown. In the Cincinnati area, *Strophomena planumbona* is largely confined to Richmondian strata of the Arnheim, Waynesville, and Liberty formations (Davis 1985).

Southern Manitoba material. Gunn Member (GSC loc. O-37130): 5 conjoined shells, 1 ventral valve.

Remarks. Only six specimens from the southern Manitoba collection are assigned to *Strophomena planumbona* on the basis of their medium-sized shell, relatively gentle geniculation, general lack of rugae except for the presence of short, weak, oblique wrinkles in the postero-lateral part of some shells (Pl. 7, figs. 1, 2). Such inconspicuous rugae are present also in some shells of the same species described from Anticosti Island (Dewing 1999) and the Hudson Bay Lowlands (Jin et al. 1997).

Strophomena vetusta James, 1874

Pl. 5, figs. 9–15; Pl. 6, figs. 1, 2

1862 *Strophomena subtenta* Billings (*non* Conrad), p. 132, fig. 109.

1874 *Streptorhynchus* (*Strophomena*) *vetusta* James, p. 241.

1912 *Strophomena vetusta* James; Foerste, p. 98, pl. 6, figs. 2a–h.

1924 *Strophomena vetusta* James; Foerste, p. 120, pl. 12, figs. 4a, b; pl. 14, fig. 5.

1926 *Strophomena vetusta* (James); Hussey, p. 179, pl. 4, figs. 12, 14.

1928 *Strophomena planocorrugata* Twenhofel, p. 194, pl. 17, figs. 4–6.

1957 *Strophomena* cf. *S. vetusta* James; Ross, p. 482, pl. 40, fig. 4.

1961 *Strophomena vetusta* James; Caster et al., pl. 8, figs. 23, 24.

1970 *Strophomena planocorrugata* Twenhofel; Macomber, p. 441, pl. 78, figs. 25–33.

1976 *Strophomena vetusta* James; Pope, text-figs. 4.1, 5.4, 6.8, 30.

1977 *Strophomena vetusta* James; Pope and Martin, pl. 8, figs. 34, 24.

1985 *Strophomena vetusta* James; Davis, pl. 8, figs. 23, 24.

1994 *Strophomena vetusta* James; Rong and Cocks, p. 668, pl. 1, fig. 8.

1995 *Strophomena vetusta* James; Dewing, p. 174, pl. 13, figs. 8, 9; pl. 14, figs. 1–9.

1997 *Strophomena vetusta* James; Jin et al., p. 25, pl. 8, figs. 1–10; pl. 9, figs. 1–8.

1999 *Strophomena vetusta* James; Dewing, p. 27, pl. 9, figs. 8, 9; pl. 10, figs. 1–9.

Types. "Upper part of Cincinnati Group, in Ohio and Indiana" (James 1874, p. 242). As discussed by Foerste (1912, p. 100), James' types probably came from the Liberty Formation or the Whitewater Formation (Richmondian), Ohio.

Southern Manitoba material. Penitentiary Member (GSC loc. O-27186 and 55-60): 2 dorsal internal moulds.

Gunn Member (GSC loc. O-37130 and C-205926): 3 conjoined shells, 3 dorsal and 4 ventral valves.

Selkirk Member (I-423): 1 dorsal valve.

Description (southern Manitoba material). Shell medium-sized to large, semicircular to subelliptical, with length ranging from 17.5–31.9 mm, width 26.9–47.6 mm, and average length/width ratio 0.65–0.85. Lateral profile weakly resupinate, with dorsal valve much more strongly convex than ventral valve. Hingeline being widest part of shell, with pointed cardinal extremities. Ventral interarea apsacline, attaining a height of 2.7 mm; delthyrium broadly triangular, covered by gently convex pseudodeltidium. Dorsal interarea much lower (<1.10 mm in height) than ventral interarea, anacline; notothyrium covered by small, arched chilidium. Entire shell covered by clearly differentiated parvicostellae, averaging 3–5 finer costellae between two adjacent coarser costellae near anterior margin. Concentric rugae irregularly spaced, being strongest toward postero-lateral margin, intersecting hingeline at near-perpendicular angle, and increasing in size (crest width) from umbonal area to cardinal extremities (Pl. 6, figs. 1). Concentric growth lines fine, evenly distributed, averaging 9 per mm.

Microscopic shell structure. Pseudopunctate, with minute pustules best developed on both sides of shell adjacent to dorsal muscle field (Pl. 5, fig. 15).

Ventral interior. Teeth small, slender (Pl. 5, fig. 12); dental plates low, thick, extending anteriorly to merge with high lateral bounding ridges of muscle field. Lateral cavities present in umbonal area, largely filled by secondary shell deposits posteriorly. Muscle field sharply defined, subcircular to subrhomboidal in outline, occupying about 34% of shell width and 42% of shell length, bearing low, relatively thick median ridge (Pl. 5, figs. 11–13); lateral bounding ridges thick and high, strongly ventro-medially onto valve floor, leaving narrow medial gap at anterior margin of muscle field; adductor muscle scars not clearly delimited.

Dorsal interior. Cardinalia small, occupying about 26% of shell width and 13% of shell length. Two lobes of cardinal process slender, discrete, with small, blade-like structure in between (Pl. 5, figs. 9, 10, 14, 15); crests of cardinal process lobes extending posteriorly beyond hingeline. Sockets deep, narrow, bounded by inner and outer socket ridges starting from hingeline and extending antero-laterally about 30°; outer socket ridges thin, short, bearing up to 8 crenulations on inner sides; inner socket ridges becoming thin and platy at distal ends, dipping posteriorly toward hingeline, and bearing 3 or 4 crenulations on outer sides. Adductor muscle field long, straight; posterior median ridge thick, strong, beginning from notothyrial platform, bifurcating anteriorly to merge with two central side septa (Pl. 5, figs. 9, 10); anterior median ridge much weaker, not continuous with posterior median septa; central side septa long, strong, slightly divergent from each other anteriorly, extending for more than half (typically two-thirds) valve length. Pair of lateral side septa starting at posterior margin of adductor muscle field, also long and relatively straight, usually somewhat shorter than central side septa.

Remarks. Shells of *Strophomena vetusta* are characterized by a large or very large size, with short but consistent postero-lateral rugae developed at approximately right angles to the hingeline. Specimens from southern Manitoba attain a maximum shell width of 47.6 mm, those of the Hudson Bay Basin 40.0 mm (Jin et al. 1997), and those of Anticosti Island 38.6 mm (Dewing 1999). Shells of *Strophomena vetusta* from southern Manitoba are virtually identical to those of *Strophomena planocorrugata* Twenhofel 1928 from the Vauréal Formation of Anticosti Island and from the Bighorn Formation of Wyoming (Macomber 1970). We agree with Dewing (1999) in treating *S. planocorrugata* as a junior synonym of *S. vetusta*, because *S. planocorrugata* was initially distinguished from *S. vetusta* only by its weak dorsal fold and somewhat smaller cardinalia, with socket ridges subparallel to the hingeline. The material of *S. vetusta* from the Churchill River Group of Hudson Bay Basin (Jin et al. 1997) show slightly longer dorsal transmuscle ridges, but it is difficult to determine the exact range of variation in this character because of limited numbers of well-preserved dorsal interiors.

Genus *Nasutimena* n. gen.

Type species. *Strophomena fluctuosa* Billings, 1860. Vauréal Formation (Richmondian), Anticosti Island (see discussion under description of species).

Etymology. From the Latin adjective, *nasutus*, large-nosed. The term *Nasutimena* describes the prominent, nose-like, antero-medial fold on the shell geniculation.

Diagnosis. Shell usually large to very large, subtriangular to subpentagonal in outline, strongly resupinate and broadly geniculate, bearing concentric to criss-cross rugae over part or entire shell. Ventral interior like *Strophomena*; dorsal interior different from *Strophomena* in having usually shorter and weaker transmuscle septa.

Remarks. *Strophomena fluctuosa* is a widespread species in Upper Ordovician rocks of North America. Assignment of the species to *Strophomena* has been mainly on the basis of its resupinate shell. The species, however, shows a number of characteristics that are distinct from *Strophomena*, as noted by Dewing (1995, 1999) who proposed to assign the species to either *Luhaia* or *Gunnarella*. Our observation of the species from Anticosti Island, the Hudson Bay Lowlands, and southern Manitoba indicates that the species possesses the following diagnostic features:

1. Continuous to irregular, criss-cross to roughly concentric rugae that are best developed in the disc area but may cover the entire shell; the rugae may or may not be interrupted by radiating costellae;
2. A strongly geniculate shell with a nasute (nose-like) antero-medial fold, especially in mature and gerontic forms, giving the shell a subtriangular outline;
3. A dorsal valve with considerably weaker dorsal transmuscle septa (Dewing 1999; Jin et al. 1997) than that of typical *Strophomena* (e.g., *S. vetusta* and *S. planumbona*).

As will be discussed under the type species, the only surviving specimen of *S. fluctuosa* from Billing's type lot from the Vauréal Formation of Anticosti Island shows distinct criss-cross rugae on the dorsal disc (see Jin et al. 1997, pl. 5, figs. 1–4), which are unlike the largely discontinuous concentric rugae in most specimens figured by Dewing (1999). Most specimens of *S. fluctuosa* described by Jin et al. (1997) from the Hudson Bay Lowlands typically show criss-cross rugae that are largely confined to the disc area. The new genus *Nasutimena* is established mainly on the basis of *S. fluctuosa*, but also includes *S. undulosa*, which has an unusually large, nasute, resupinate shell, with very strong criss-cross rugae covering nearly the entire shell.

Species included:

Strophomena fluctuosa Billings, 1860 (see description of species).

Strophomena undulosa Roy, 1941 (see description of species).

Age and distribution: Maysvillian–Richmondian (Ashgill); North America.

Nasutimena fluctuosa (Billings, 1860)

Pl. 6, figs. 9–19; Pl. 7, figs. 5–12; Pl. 20, figs. 1–7; Pl. 21, figs. 5, 6; Figs. 16, 17

1860 *Strophomena fluctuosa* Billings, p. 57, fig. 6.

1862 *Strophomena fluctuosa* Billings; Billings, p. 123, figs. 102a, b.

1863 *Strophomena fluctuosa* Billings; Logan, p. 221, fig. 207A.

1892 *Strophomena fluctuosa* Billings; Hall and Clarke, p. 251, pl. 11A, figs. 4, 5.

1924 *Strophomena fluctuosa* Billings; Foerste, p. 119, pl. 12, figs. 8a, b.

1928 *Strophomena fluctuosa* Billings; Twenhofel, p. 193, pl. 22, figs. 3–5.

1928 *Strophomena fluctuosa* Billings; Troedsson, p. 93, pl. 22, fig. 13.

1970 *Strophomena fluctuosa* Billings; Macomber, p. 440; pl. 78, figs. 16–24.

1972 *Strophomena fluctuosa* Billings; Bolton, p. 22, pl. 1, fig. 9.

1981 *Strophomena fluctuosa* Billings; Bolton, p. 50, pl. 3, fig. 4.

1995 *Luhaia fluctuosa* (Billings); Dewing, p. 139, pl. 10, figs. 2, 4–8; pl. 11, figs. 1–13.

1997 *Strophomena fluctuosa* Billings; Jin et al., p. 26, pl. 5, figs. 1–16; pl. 6, figs. 1–7.

1999 *Gunnarella fluctuosa* (Billings); Dewing, p. 20, pl. 6, figs. 2, 4–8; pl. 7, figs. 1–13.

Types. The original type specimen of *Strophomena fluctuosa* was reported as missing by Wilson (1945). The only published hypotype (GSC 2017) of this species is from the Vauréal Formation, Carlton Point, Anticosti, and was illustrated by Twenhofel (1928, pl. 22, fig. 5) and Jin et al. (1997, pl. 5, figs. 1–4). Billings (1860, 1862) originally reported the species from the "Trenton limestone, City of Ottawa, rare; more common in the Hudson River Group [= Vauréal Formation], Anticosti." Specimen GSC 2017, however, partly matches Billings's (1860) figure, especially the sinuous lateral margin on the right side (in dorsal view) and the presence of rugae. Billings (1860, p. 58) originally described the shell ornamentation as follows:

> "In most of the specimens, the whole of the upper half of the shell is covered with short undulating wrinkles, which sometimes have a concentric arrangement and often form concentric rows converging from the hinge-line towards the centre of the shell, cross each other. The specimens from the Trenton limestone [of Ottawa area] are usually without these undulations, but in those from the Hudson River group [Vauréal Formation of Anticosti Island] this character is prominently exhibited."

The strongly wrinkled shell illustrated by Billings indicates that the type locality is undoubtedly Anticosti Island, not the Ottawa area. The perfect shell illustrated in Billings (1860) may have been the result of the artist's embellishment, as was common in his early publications. In the same paper, for example, Billings (1860) also described another new species, *Strophomena hecuba*, accompanied with a drawing of a perfect shell (holotype by monotypy), which actually lacks a large part around one of the ears (see Twenhofel 1928, pl. 23, fig. 3). The specimen GSC 2017 may have been miscatalogued and can be reasonably regarded as the holotype by monotypy. The designation of a neotype by Dewing (1999), therefore, is unnecessary.

Southern Manitoba material. Penitentiary Member (GSC loc. O-27186 and 55-60): 20 dorsal internal and 1 external, 5 ventral internal and 10 external moulds, and 3 complete internal moulds.

Gunn Member (GSC loc. O-37130, C-205926, C-205930, C-205931): 17 conjoined shells, 23 dorsal and 13 ventral valves.

Selkirk Member (MMMN I-2153, I-2853): 4 individuals with conjoined valves and 1 dorsal valve.

Dog Head Member (MMMN I-2416): 1 dorsal valve.

Description (southern Manitoba material). Shell usually large (Fig. 16), subtriangular in outline of adult forms, with average length of 18.0 mm (maximum 31.8 mm) and width 24.0 mm (maximum 40.5 mm). Lateral profile resupinate, with strong geniculation beginning at 11–17 mm from apex and bending ventrally at 120–150°; antero-medial part of trail marked by nasute (nose-like) fold in dorsal valve and corresponding sulcus in ventral valve. Hingeline long, usually being widest part of shell, extending into conspicuous ears in some specimens. Ventral interarea attaining up to 2.2 mm in height, catacline (common) to apsacline (rare); delthyrium covered by strongly convex pseudodeltidium. Dorsal interarea lower than ventral interarea, invariably anacline; notothyrium covered by convex chilidium. Entire shell covered with well-differentiated parvicostellae; major costellae increasing anteriorly by intercalation and minor costellae increasing more commonly through bifurcation, with two to five minor costellae between two adjacent major costellae toward shell margin. Concentric growth lines fine, present over entire shell surface but better developed in anterior half of shell, averaging 10 per 1 mm. Concentric rugae variably developed, strongest in disc area, rarely present in trail, ranging from disrupted by major costellae to locally continuous (superimposed on costellae), even within single shell (Pl. 6, figs. 11, 14; Pl. 7, fig. 7).

Fig. 16. Shell dimensions of *Nasutimena fluctuosa* (Billings, 1860), sample from GSC loc. O-37130, Gunn Member, Stony Mountain Formation. Note variability in length/width ratio.

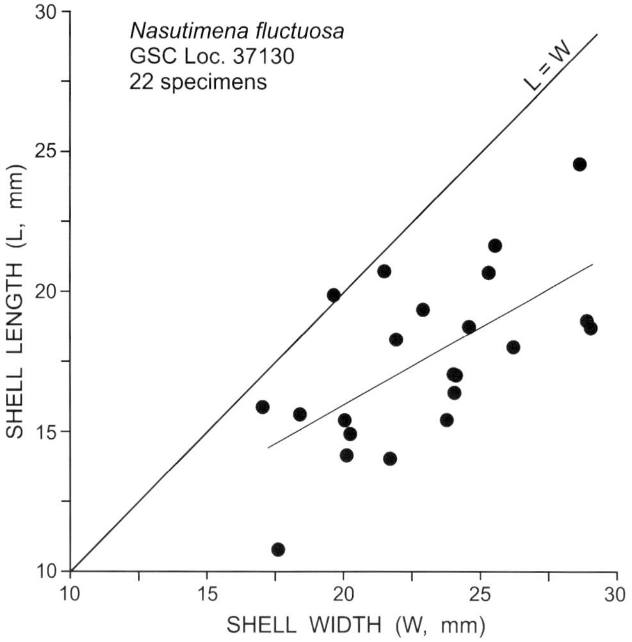

Statistics of shell measurements (mm):

GSC loc. O-37130	*L*	*W*	*L/W*
AVG	18.0	24.0	0.75
STD	4.6	5.6	0.11
MIN	7.5	12.8	0.58
MAX	31.8	40.5	1.01

Microscopic shell structure. Shell finely pseudopunctate, averaging 11–13 per mm in cross sections (Fig. 17).

Ventral interior. Teeth strong, massive (Fig. 17), each bearing 7–9 weak crenulations; dental plates short, stubby, extending antero-medially to form surrounding ridges of circular or subcircular muscle field. Pair of diductor scars divided by strong median ridge (myophragm), tapering and terminating slightly anterior of muscle field. Adductor scars suboval in outline, located in medial part of muscle field, enclosed laterally by diductor scars (Pl. 20, figs. 1, 3). Peripheral ridge (rim) clearly developed (Pl. 6, fig. 17), starting near midpoint on each side of hingeline, bending antero-laterally and then converging medially along margin of visceral disc. Anterior medial portion of trail marked by prominent crenulations (Pl. 6, fig. 17).

Dorsal interior. Notothyrial platform notably elevated, with posterior part almost fully occupied by robust cardinal process. Two lobes of cardinal process fused at base, protruding ventrally into delthyrial cavity (Fig. 17, 1.1 mm from apex). Sockets deep, wide, bounded medially by low inner socket ridges. Median ridge (myophragm) short, strong, originating from notothyrial platform and extending anteriorly for about one-fourth of valve length (Fig. 17); transmuscle septa variously developed (Pl. 20, figs. 2, 4–7). Peripheral ridge well developed, corresponding to that of ventral valve.

Remarks. The shells of *Nasutimena fluctuosa* from southern Manitoba show a wide range of variation in shell outline, which is reflected partly in the plot of their length/width ratios (Fig. 16). Considerable variation is observed also in the development of rugae,

Fig. 17. Transverse serial sections of *Nasutimena fluctuosa* (Billings, 1860), hypotype, GSC 117794 (mature shell: *L* = 19.3 mm, *W* = 22.9 mm, *T* = 8.34 mm), GSC loc. O-37130, Gunn Member, Stony Mountain Formation.

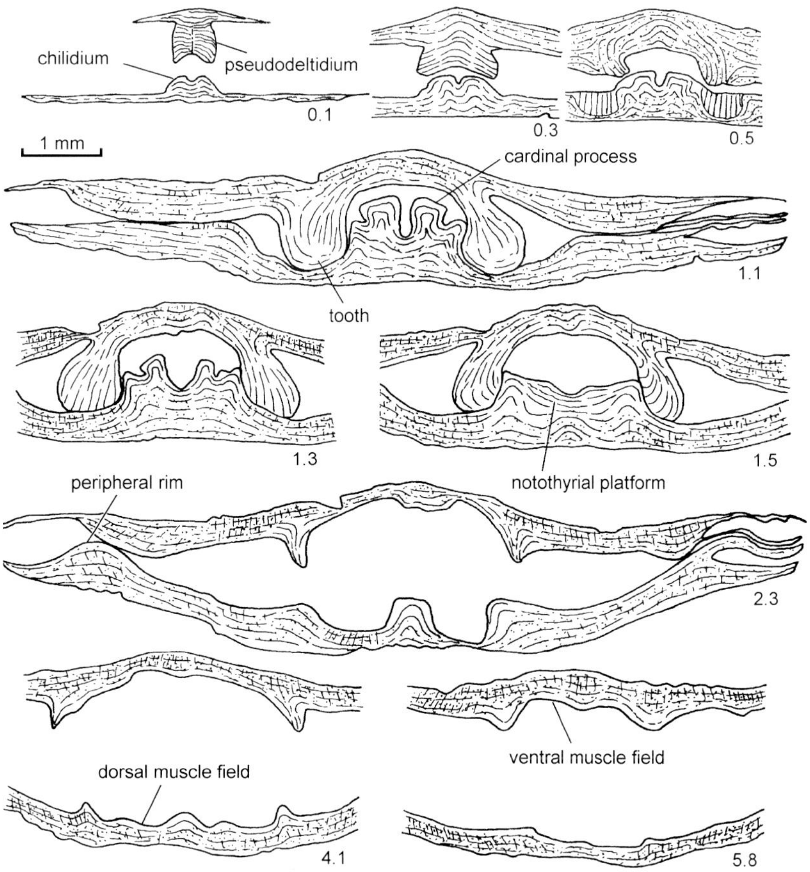

which range from crudely concentric to criss-cross, from weak to strong, and may be disrupted by, or superimposed on, radiating costellae. Shells of *Nasutimena fluctuosa* from the Bighorn Formation of Wyoming are virtually identical to the southern Manitoba material except that the Wyoming shells appear to have a poorly developed chilidium (see Macomber 1970, pl. 78, fig. 20), although this may have been the result of preservation. Specimens of the same species from the Caution Creek Formation of the Hudson Bay Basin (Jin et al. 1997) and the Vauréal Formation of Anticosti Island (Dewing 1999) seem to have stronger concentric rugae and a high trail. Among all the occurrences, the Hudson Bay material appears to show the best developed criss-cross rugae, as initially noted also by Billings (1860) for the Anticosti shells.

Dewing (1999) treated *Strophomena occidentalis* Foerste, 1912, as a junior synonym of *N. fluctuosa*. Shells of *Strophomena occidentalis* from Iowa (Wang 1949), Minnesota (Winchell and Schuchert 1893), and the Hudson Bay Lowlands (Jin et al. 1997) typically have a smaller shell with relatively fine concentric (not criss-cross) rugae on the disc. Pending more detailed examination of its internal structures, *S. occidentalis* is provisionally regarded as an independent species.

Nasutimena undulosa (Roy, 1941)

Pl. 7, figs. 13, 14; Pl. 8, figs. 1–4

1928 *Strophomena fluctuosa* Billings; Troedsson, p. 93, pl. 22, fig. 13.

1941 *Strophomena undulosa* Roy, p. 98, fig. 63.

?1977 *Strophomena undulosa* (Roy); Bolton, p. 58, pl. 10, figs. 6, 13.

1997 *Strophomena undulosa* Roy; Jin et al., p. 6, figs. 8–10; pl. 7, figs. 1–4.

1998 *Strophomena undulosa* Roy; Jin and Copper, p. 445, figs. 3–5.

Types. The holotype designated by Roy (1941), YPM 28273, came from Silliman's Fossil Mount (upper Maysvillian), Baffin Island.

Southern Manitoba material. Two hypotypes (GSC 109024 and GSC 109025) from the Selkirk Member, Red River Formation (Maysvillian), Garson Quarry (GSC loc. C-205935), southern Manitoba.

Remarks. Roy (1941) established the species *Strophomena undulosa* on the basis of medium-sized shells (18.1 mm long and 22.8 mm wide) having criss-cross rugae over the entire surface of the shell. The species is assigned to the new genus because of its triangular shell outline and unique development of rugae.

The only two unusually large shells from southern Manitoba are assigned to *Nasutimena undulosa* mainly on the basis of their strong, sharply defined criss-cross rugae developed over the entire shell (i.e., both on the disc and the trail). The sharply developed, *Kulumbella*-type divaricate ribbing (Jin and Copper 1998) in the southern Manitoba shells is distinct in strophomenoids.

Genus *Holtedahlina* Foerste, 1924

Type species. *Leptaena sulcata* de Verneuil, 1848. Waynesville Formation (Richmondian), Indiana, U.S.A.

Holtedahlina paraprostrata n. sp.

Pl. 8, figs. 5–8

Types. Three specimens are selected as types: holotype, MMMN I-2543b (Pl. 8, fig. 7), paratypes, MMMN I-2543a and MMMN I-2543c (Pl. 8, figs. 5, 6, and 8). All from the Fort Garry Member, Red River Formation, west bank of Red River, southern Manitoba.

Etymology. From the Latin adjective, *prostratus*, lying low. The term *paraprostrata*, a feminine adjective, denotes the nearly flat to weakly biconvex shell of the new species.

Additional material. Fort Garry Member (loc. MMMN I-2543): 1 ventral internal and 1 external moulds.

Diagnosis. Relatively small, weakly biconvex shells of *Holtedahlina* with largely open delthyrium and notothyrium, subparallel dental plates, and small cardinal process extended posteriorly of hingeline.

Description. Shell relatively small, barely reaching medium size, semicircular in outline, ranging from 8.8–10.0 mm long and 13.2–16.3 mm wide. Lateral profile gently biconvex, with ventral valve slightly deeper than dorsal valve. Cardinal extremities right-angled or somewhat rounded. Ventral interarea low (0.36 mm in height), apsacline; delthyrium covered posteriorly by pseudodeltidium; dorsal interarea attaining 0.31 mm in height, anacline; notothyrium covered posteriorly by small chilidium. Entire shell covered by coarse, equal-sized costellae, increasing anteriorly several times through intercalation and averaging 3 per 1 mm near anterior margin. Concentric growth lines present, interrupted by several concentric, unevenly spaced growth lamellae in anterior half of shell.

Ventral interior. Teeth small; dental plates thin, low, diverging from each other antero-laterally at about 85°, becoming parallel to each other in their distal portions and extending into lateral bounding ridges of muscle field. Muscle field suboval in outline, open anteriorly (Pl. 8, fig. 5), occupying one-third of shell length and one-seventh of shell width.

Dorsal interior. Cardinalia small, delicate, occupying about one-tenth of shell length and one-fourth of shell width. Notothyrial platform absent. Cardinal process protruding posteriorly beyond hingeline, with two small, discrete lobes. Socket ridges short, thin, diverging antero-laterally at about 120°, medially continuous with base of cardinal process. Adductor muscle field shallow, occupying less than 17% of shell length or width, divided by short, weak median ridge; transmuscle ridges not developed.

Remarks. There are four strophomenine genera with biconvex shells: *Infurca* Percival, 1979, *Esillia* Nikitin and Popov 1985, *Trotlandella* Neuman (in Neuman and Bruton) 1974, and *Holtedahlina* Foerste 1924, which can be distinguished from one another largely on the basis of internal structures (Zhan and Cocks 1998). *Teratelasma* Cooper, 1956, is a strophomenoid with gently biconvex lateral profile and a pair of small cardinal process lobes, but it differs from the above four genera in its strongly developed and uniquely arranged dorsal median septum and transmuscle ridges.

The new species is the first record of *Holtedahlina* from the Red River Formation of southern Manitoba. It can be distinguished from other congeneric species in its extremely thin shell, largely open delthyrium and notothyrium, subparallel dental plates, and a small cardinal process that extends posteriorly of the hingeline. *Holtedahlina* sp. from the Fort Atkinson Member of the Maquoketa Formation of Iowa (Wang 1949) is similar to the new species in its equal-sized costellae and subparallel dental plates, but differs in its much larger and thicker shell, higher dental plates, and well-developed pseudodeltidium. *Holtedahlina sinica* Zhan and Cocks (1998, p. 44, pl. 5, figs. 18–21; pl. 6, figs. 1, 2), from the Changwu Formation (mid-Ashgill) of South China, also resembles the new species in its thin shell and largely open delthyrium, but differs in its high and centrally dipping dental plates and larger and more robust cardinalia.

Genus *Tetraphalerella* Wang, 1949

Type species. Tetraphalerella cooperi Wang. Upper Elgin Member (early Maysvillian), Maquoketa Formation, Iowa.

Remarks. Wang (1949) initially distinguished *Tetraphalerella* from *Strophomena* by its generally larger shell, more strongly transverse outline, more prominent and well-arranged pseudopunctae, and much thinner socket ridges that are undercut and dip posteriorly (see also Jin et al. 1997). Dewing (1999) proposed a new family, the Tetraphalerellidae, largely on the basis of the microscopic shell structure of *Tetraphalerella*, which has uncored pseudopunctae arranged in regular radial rows, particularly in the anterior part of the disc area. The pseudopunctae of *Strophomena*, in comparison, have a smooth teleolate core and are usually arranged irregularly over the entire shell. Dewing (1999) further proposed a closer affinity of *Tetraphalerella* to *Coolinia* than to *Strophomena* in terms of their denticulation and cardinal process. Shells of *Tetraphalerella* from both the Hudson Bay Lowlands (Jin et al. 1997) and southern Manitoba confirm Dewing's observation on the characteristic development of pseudopunctae of the genus. The consistency of the size and arrangement of pseudopunctae, however, need further investigation. Some shells of *Tetraphalerella* from the Hudson Bay Lowlands, for example, may have pseudopunctae regularly arranged in radial rows in the disc area but randomly arranged in the marginal, geniculate part of the same valve (Jin et al. 1997). Pending a broader survey of shell microstructures in the strophomenoids and a thorough evaluation of the taxonomic importance of the pseudopunctae, the genus *Tetraphalerella* is retained provisionally here in the Strophomenidae.

In the revised brachiopod volumes of the *Treatise on Invertebrate Paleontology*, Cocks and Rong (2000) treated *Tetraphalerella* as a subgenus of *Strophomena*, without taking microscopic shell structures into consideration. Because of the opposing views of Dewing (1999) and Cocks and Rong (2000), *Tetraphalerella* and *Strophomena* are retained herein as independent genera.

Tetraphalerella neglecta (James, 1881)

Pl. 9, figs. 3–7

1881 *Streptorhynchus neglectum* James, p. 41.

1912 *Strophomena neglecta* (James); Foerste, pp. 90–95, pl. 5, figs. 1, 3; pl. 7, fig. 5; pl. 9, figs. 1, 10; pl. 11, fig. 10.

1924 *Strophomena neglecta* (James); Foerste, p. 121, pl. 13, figs. 5a, b.

1949 *Tetraphalerella neglecta* (James); Wang, p. 30, pl. 9G, figs. 1–4.

1988 *Tetraphalerella neglecta* (James); Howe, p. 216, fig. 6.29.

Southern Manitoba material. Selkirk Member (MMMN I-2855): 2 dorsal and 1 ventral valves.

Cat Head Member (UM P198); 4 dorsal internal, 3 dorsal external, and 2 ventral external moulds.

Description (southern Manitoba material). Shell usually medium-sized, semicircular to semielliptical, ranging from 17.7–26.4 mm in length and 25.7–37.9 mm in width, with an average length/width ratio of 0.7. Lateral profile gently resupinate, with visceral disc gently and evenly convexo–concave, and geniculation at about 110°. Hingeline straight, being widest part of shell; cardinal extremities extending into inconspicuous ears. Ventral interarea apsacline, attaining height of 1.4 mm; delthyrium and pseudodeltidium obscured by matrix. Dorsal interarea slightly lower than ventral interarea, anacline, attaining height of

1.2 mm; chilidium small, convex, covering posterior part of cardinal process. Entire shell covered by unequal, closely spaced parvicostellae (averaging 8 per 1 mm in medial part of shell); coarser costellae unevenly spaced, increasing anteriorly through intercalation. Concentric growth lines fine, evenly spaced over entire shell surface.

Ventral interior. (Not studied because of insufficient material.)

Dorsal interior. Cardinalia small, occupying about 6% of shell length and 21% of shell width; cardinal process lobes converging toward their bases, with each lobe bearing longitudinal depression on posterior surface of myophore. Socket ridges thin, plate-like, undercut anteriorly, continuous with base of cardinal process medially, extending antero-laterally and then postero-laterally toward hingeline, bounding shallow sockets. Muscle field shallow, subcircular in outline, occupying less than one-quarter of shell width and one-third shell length, divided by strong median ridge originating from low notothyrial platform; transmuscle septa lacking.

Remarks. Compared to *Tetraphalerella planodorsata* (Winchell and Schuchert, 1892) from the Elgin Member of the Maquoketa Formation of Iowa (Wang 1949) and the Surprise Creek Formation of Hudson Bay Lowlands (Jin et al. 1997), *Tetraphalerella neglecta* has a notably smaller shell with a less sharp geniculation.

Tetraphalerella churchillensis Jin, Caldwell, and Norford, 1997

Pl. 9, figs. 1, 2, 8, 9

1997 *Tetraphalerella churchillensis* Jin , Caldwell, and Norford, p. 30, pl. 10, figs. 7–9; pl. 11, figs. 4–9; pl. 12, figs. 1–3.

Types. The original types of Jin et al. (1997) came from the Surprise Creek Formation, Maysvillian, Churchill River, Hudson Bay Lowlands.

Southern Manitoba material. Selkirk Member (MMMN I-185, I-1678): 3 ventral valves, 2 dorsal internal and 2 dorsal external, and 1 ventral external moulds.

Remarks. Shells of *Tetraphalerella churchillensis* from southern Manitoba are virtually identical to those from the Hudson Bay Lowlands in their unusually large size and low convexity. *Tetraphalerella churchillensis* ranks among the three largest-shelled species among the Red River brachiopods and may well be the largest Ordovician strophomenids ever recorded. The largest, measurable specimens in the present collection attain a maximum length of 40 mm and width of 70 mm. The subcircular ventral muscle field measures a maximum of 12.8 mm in length and 14.3 mm in width. In their description of the Hudson Bay material, Jin et al. (1997) illustrated a specimen from Garson Quarry (GSC loc. C-205935), southern Manitoba, and provided detailed description and discussion of the species.

Family RAFINESQUINIDAE Schuchert, 1893

Genus *Kjaerina* Bancroft, 1929

Type species. *Kjaerina typa* Bancroft, 1929. Glynboro Member, Cheney Longville Formation (Longvillian, Caradoc), Shropshire, England.

Kjaerina hartae Jin, Caldwell, and Norford, 1995

Pl. 13, figs. 1–14

1928 *Rafinesquina* sp. (cf. *R. deltoidea* Conrad) Troedsson, p. 90, pl. 22, figs. 9a, 9b.

1995 *Kjaerina hartae* Jin, Caldwell, and Norford, p. 1260, pl. 1, figs. 3–11; pl. 2, figs. 1–9.

1997 *Kjaerina hartae* Jin, Caldwell, and Norford, p. 33, pl. 12, figs. 4–6; pl. 16, figs. 9–11; pl. 17, figs. 1–12; pl. 18, figs. 1–4; pl. 22, figs. 4–8.

Types. Holotype GSC 109077, selected by Jin et al. (1995), came from the Selkirk Member, Red River Formation, Garson Quarry (GSC loc. C-205935), southern Manitoba.

Additional material examined. Selkirk Member (MMMN I-94, I-686, I-2152, I-2852): 6 broken individuals with conjoined valves, and 13 ventral valves.

Remarks. In terms of maximum size, *Kjaerina hartae* is the third largest-shelled species among the Red River brachiopods, attaining 30–40 mm in length and 40–55 mm in width. Other characteristics of the species include the strongly transverse outline and well-developed, uneven, concentric rugae. In addition to the detailed description of the species by Jin et al. (1995, 1997), it is observed in this study that the costellae on the trail are obviously more densely spaced (averaging 4–5 per mm) than those on the visceral disc (averaging 3 per mm), and that all the new costellae increase anteriorly by bifurcation. One of the specimens from the Cape Calhoun Formation of northern Greenland, described by Troedsson (1928, pl. 22, figs. 9a and 9b) as *Rafinesquina* sp. (cf. *R. deltoidea* Conrad), measures 33.2 mm in length and 42.5 mm in width, and its outline and ornament are virtually identical to those of *Kjaerina hartae* from southern Manitoba and the Hudson Bay Lowlands.

Genus *Megamyonia* Wang, 1949

Type species. *Megamyonia knighti* Wang, 1949. Fort Atkinson Member (late Maysvillian), Maquoketa Formation, Iowa.

Remarks. In establishing the genus *Megamyonia*, Wang (1949) emphasized that "the most important external difference between *Leptaena* and *Megamyonia* is the lack of concentric wrinkles in the latter genus." In the southern Manitoba collection of *Megamyonia*, some shells show very weak concentric rugae on the visceral disc. Other diagnostic features of *Megamyonia*, noted for the Iowa material, are characteristic of the southern Manitoba representatives, including the inflated medial costa on the ventral valve, the unusually large, flabellate to subcircular ventral muscle field, and the lack of transmuscle septa in the dorsal valve.

Megamyonia nitens (Billings, 1860)

Pl. 14, figs. 1–13; Pl. 22, figs. 1, 2; Figs. 18, 19

1860 *Strophomena nitens* Billings, p. 53, fig. 1.

1862 *Strophomena nitens* Billings; Billings, p. 118, fig. 97.

1917 *Strophomena (?) tetrastriata* Parks, p. 16, pl. 4, fig. 16.

1928 *Leptaena? nitens* (Billings); Twenhofel, p. 186, pl. 17, fig. 19; pl. 18, figs. 13, 14.

1956 *Megamyonia nitens* (Billings); Stearn, p. 101, pl. 11, figs. 9–11.

1957 *Megamyonia* cf. *M. unicostata* (Meek and Worthen); Zhan, p. 483, pl. 40, figs. 23–25.

1957 *Megamyonia* cf. *M. ceres* (Billings); Ross, p. 484, pl. 40, figs. 15, 17, 19–21.

1957 *Megamyonia* aff. *M. nitens* (Billings); Ross, p. 484, pl. 41, figs. 3, 4, 7.

1960 *Megamyonia* cf. *nitens* (Billings); Brindle, pl. 4, fig. 13.

1970 *Megamyonia nitens* (Billings); Macomber, p. 442, pl. 79, figs. 9–25.

1997 *Megamyonia nitens* (Billings); Jin et al., p. 31, pl. 13, figs. 6–14; pl. 14, figs. 1–8.

1999 *Megamyonia nitens* (Billings); Dewing, p. 38, pl. 14, figs. 4–14.

Type. The lectotype, GSC 2019, selected by Dewing (1999) and photographically illustrated in Twenhofel (1928, pl, 18, fig. 14), came from the exposed lower part of the Vauréal Formation (Richmondian), Carleton Point, Anticosti Island, Québec.

Southern Manitoba material. Penitentiary Member (GSC loc. O-27186): 1 dorsal external mould.

Gunn Member (GSC loc. O-27185, O-37130, and C-205931): 45 conjoined shells, 1 dorsal and 77 ventral valves.

Selkirk Member, Garson Quarry (GSC loc. C-205935): 2 conjoined shells and 4 ventral valves.

Description (southern Manitoba material). Shell medium-sized, semicircular to transversely subelliptical, with average length of 12.3 mm (maximum 18.5 mm), width 18.3 mm (maximum 23.3 mm), and average length/width ratio of 0.67 (Fig. 18); lateral profile concavo-convex, with variously developed geniculation ranging from virtually absent to short and gentle to long and sharp (longer than visceral disc); deepest part of shell located centrally. Hingeline long, being widest part of shell, with cardinal extremities extending into small ears in some specimens. Ventral umbo slightly swollen; interarea anacline; delthyrium covered posteriorly by small, strongly convex pseudodeltidium. Dorsal interarea much lower than ventral interarea, hypercline; notothyrium covered by well-developed chilidium. Shell ornamented by weak parvicostellae with prominent ventral median costella; usually 3–5

Fig. 18. Shell dimensions of *Megamyonia nitens* (Billings, 1860), sample from GSC loc. O-37130, Gunn Member, Stony Mountain Formation. Note wide range of variation in length/width ratio.

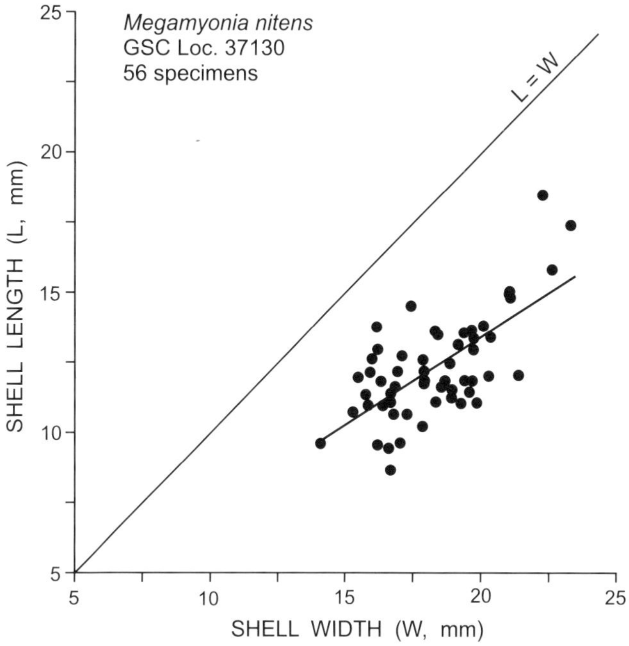

finer costellae between two coarser ones; all costellae increasing anteriorly through intercalation. Concentric growth lines well developed and evenly spaced on entire shell surface; concentric rugae weakly developed on visceral disc, particularly on both sides of some shells.

Statistics of shell measurements (mm):

GSC loc. O-37130	*L*	*W*	*L/W*
AVG	12.3	18.3	0.67
STD	1.9	2	0.08
MIN	8.7	14.1	0.52
MAX	18.5	23.3	0.85

Microscopic shell structure. Pseudopunctae large, arranged regularly along interspaces of costellae (Pl. 14, figs. 8, 9).

Ventral interior. Teeth small, wedge-shaped, supported directly by the thickened lateral shell wall; dental plates absent. Muscle field deeply impressed, divided by low, thick median ridge (myophragm); lateral bounding ridges weakly developed posteriorly, disappearing rapidly in anterior portion of muscle field;

adductor muscle scars small, suboval in outline, located in postero-medial part of muscle field and enclosed completely by diductor scars; diductor muscle scars large, flabellate in outline, more than twice as long as adductor scars; transmuscle ridges poorly developed or absent.

Dorsal interior. Cardinalia small; cardinal process bilobed, sitting on posterior part of notothyrial platform; lobes of cardinal process high, extending ventrally and posteriorly, and converging at their bases (Fig. 19, 1.0 mm from apex). Sockets narrow, deep, with short and weak socket ridges. Adductor muscle field posteriorly divided by median ridge, anteriorly pustular.

Remarks. Rong and Cocks (1994) regarded the loss of dental plates as an important evolutionary trend in strophomenoids, but did not note the absence of dental plates in *Megamyonia*. According to their observation, lack of dental plates tends to be associated with development of denticles along the hingeline or at least some crenulations on the teeth and in the sockets. This does not seem to be the case in *Megamyonia nitens*, which lacks dental plates, denticles, or dental crenulations. The wide range of variation in external morphology has been discussed in detail (e.g., Macomber 1970; Jin et al. 1997). Serial sectioning in this study concurs with that of Dewing (1999) in showing the two cardinal process lobes converging onto the notothyrial platform, although the degree of convergence appears to be greater in the southern Manitoba specimen than in the Anticosti shell sectioned by Dewing.

Megamyonia nitens can be distinguished from the type species, *Megamyonia knighti*, by its more strongly transverse shell, generally stronger geniculation, and the presence of weak concentric rugae. The southern Manitoba shells of *M. nitens* are very similar to *Megamyonia unicostata* from the Maquoketa Formation of Iowa (Wang 1949), but the Iowa species has equal-sized costellae and lacks concentric rugae. Judging from the external morphology, the shells of *Megamyonia* aff. *M. nitens* (Ross 1957) from the uppermost Bighorn Dolomite of Wyoming have closer affinity to *Oepikina limbrata* than to *Megamyonia nitens*.

Fig. 19. Transverse serial sections of *Megamyonia nitens* (Billings, 1860), hypotype, GSC 117795 (mature shell: L = 13.4 mm, W = 17.3 mm, T = 6.4 mm), GSC loc. O-37130, Gunn Member, Stony Mountain Formation.

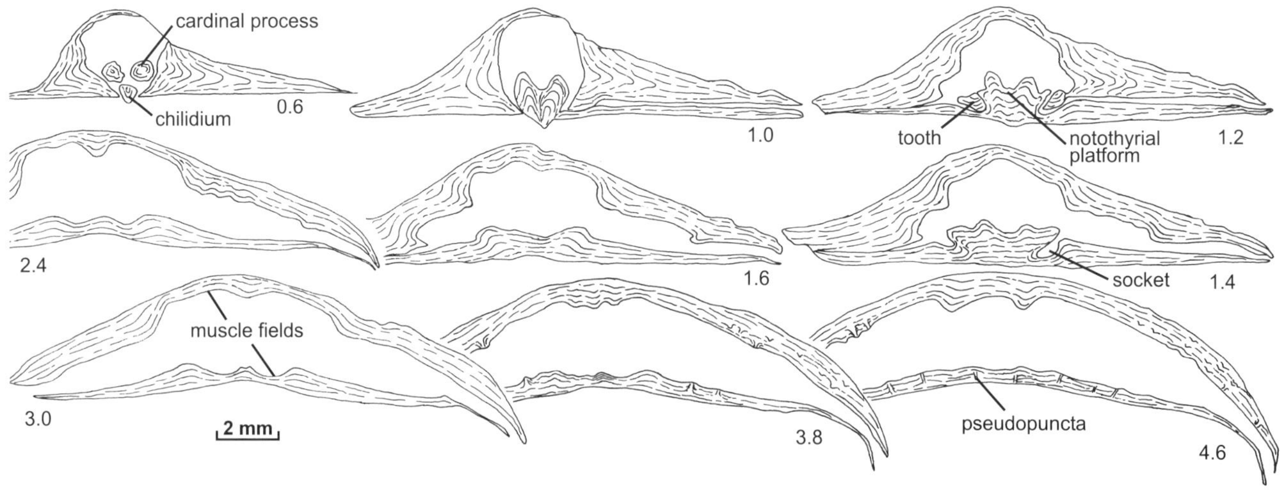

Family OEPIKINIDAE Sokolskaya, 1960

Genus *Oepikina* Salmon, 1942

Type species. *Oepikina septata* Salmon, 1942. Lebanon Formation (Mohawkian), Tennessee.

Remarks. Rong and Cocks (1994) rejected the Oepikinids as an independent family or subfamily mainly on the basis of their type-A cardinal process. Jin et al. (1997) argued for the validity of the Oepikinidae, which was further supported by the cladistic analysis of Kaplan et al. (1999), who showed that the Oepikininae constitute a distinct group within the Strophomenoidea. Following Jin et al. (1997) and Dewing (1999), the family Oepikinidae is retained to include those members having a concavo-convex, finely pseudopunctate shell with a large, unbounded ventral muscle field.

Oepikina lata (Whiteaves, 1896)

Pl. 10, figs. 1–13; Pl. 11, figs. 1–10

1896 *Rafinesquina lata* Whiteaves (in part), p. 392 (no illustrations).

1897 *Rafinesquina lata*; Whiteaves (in part), p. 172, pl. 19, figs. 2, 2a, 4, 5.

1995 *Oepikina lata* (Whiteaves), Jin et al., p. 1264, pl. 1, figs. 1, 2; pl. 3, figs. 1–8.

Types. Holotype by monotypy, GSC 4391, Selkirk Member (Maysvillian), Red River Formation, Lower Fort Garry, southern Manitoba.

Additional material examined. Selkirk Member (MMMN I-377, I-1654, I-1871, I-2150, I-2833, and I-2847): 5 conjoined shells, 1 dorsal and 12 ventral valves.

Remarks. A detailed redescription of the species can be found in Jin et al. (1995). The shells of *Oepikina lata* are unusually large and thick, attaining a maximum length of 55 mm and width of nearly 70 mm. The only addition to the description of Jin et al. (1997) is that the dorsal valve is somewhat thinner in the visceral disc area than in the marginal part, thus forming a subcircular peripheral rim that originates from the midpoint of each side of the hingeline.

Oepikina limbrata Wang, 1949

Pl. 12, figs. 1–21; Fig. 20

1943 *Rafinesquina ceres* (Billings); Okulitch, p. 62.

1949 *Öpikina limbrata* Wang, p. 22, pl. 6B, figs. 1–7.

1963 *Megamyonia ceres* (Billings); Nelson, pp. 15, 16.

1964 *Megamyonia ceres* (Billings); Nelson, pp. 23–25.

1970 *Oepikina* cf. *O. pergibbosa* (Foerste), Macomber, p. 441, pl. 79, figs. 1–8.

1997 *Oepikina limbrata* Wang; Jin et al., p. 36, pl. 20, figs. 1–14; pl. 21, figs. 1–8.

Types. The type lot illustrated by Wang (1949) is from the Lower Maquoketa Formation (Maysvillian), Clermont, Iowa.

Southern Manitoba material. Penitentiary Member (GSC loc. O-27186 and 55-60): 6 dorsal internal,

8 dorsal external, 29 ventral internal, and 1 ventral external moulds, and 2 complete internal moulds.

Gunn Member (GSC loc. O-27185, O-37130, C-205926, C-205928, C-205930, and C-205931): 39 conjoined shells, 133 dorsal and 135 ventral valves.

Description (southern Manitoba material). Shell usually medium-sized to relatively large, semicircular in outline, with average length of 21.3 mm (maximum 29.8 mm), width 24.8 mm (maximum 34.61 mm), and length/width ratio of 0.86 (Fig. 20). Lateral profile strongly concavo–convex; dorsal visceral disc gently concave, becoming variously geniculate marginally; ventral valve evenly convex, with deepest part at about one-third of shell length from apex. Ventral umbo swollen, with minute, erect beak (Pl. 12, figs. 2, 15, 17). Hingeline usually being widest part of shell. Ventral interarea anacline, up to 2.3 mm in height; delthyrium covered by narrow, arched pseudodeltidium; dorsal interarea much lower than ventral interarea, averaging 1.3 mm in height, hypercline; notothyrium covered by strongly convex chilidium. Foramen small, mesothyridid. Shell unequally parvicostellate, with costellae increasing anteriorly through intercalation, reaching 3–6 finer costellae (commonly 4 or 5) between two coarser ones in ventral valve, and 3–5 (usually 3) between two coarser costellae in dorsal valve. Weak rugae present in postero-lateral parts of both valves in some shells. Concentric growth lines well developed and evenly spaced over entire shell surface. Strong concentric lamellae usually present near anterior margin of some relatively large shells.

Statistics of shell measurements (*mm*):

GSC loc. O-37130	*L*	*W*	*T*	*L/W*	*T/W*
AVG	21.3	24.8	11.8	0.86	0.47
STD	4.1	4.13	3	0.07	0.08
MIN	13.3	15.8	7.1	0.75	0.34
MAX	29.8	34.6	17.2	1.02	0.69

Microscopic shell structure. Fine and dense pseudopunctae randomly arranged.

Ventral interior. Teeth robust, bearing 4–6 (commonly 6) crenulations on inner surfaces; dental plates thick, extending antero-laterally into bounding ridges of muscle field. Muscle field large, subcircular in outline, deeply impressed, corresponding to deepest part of shell; adductor muscle scars small, slender, slightly elevated above floor of muscle field, divided by posterior median ridge, and completely enclosed by much larger diductor scars. Median ridge low, wide, starting from delthyrial cavity, and ending near anterior part of peripheral rim. Peripheral rim originating from lateral sides of teeth, bulging anteriorly to enclose circular area (Pl. 12, figs. 7, 8).

Fig. 20. Shell dimensions of *Oepikina limbrata* Wang, 1949, sample from GSC loc. O-37130, Gunn Member, Stony Mountain Formation.

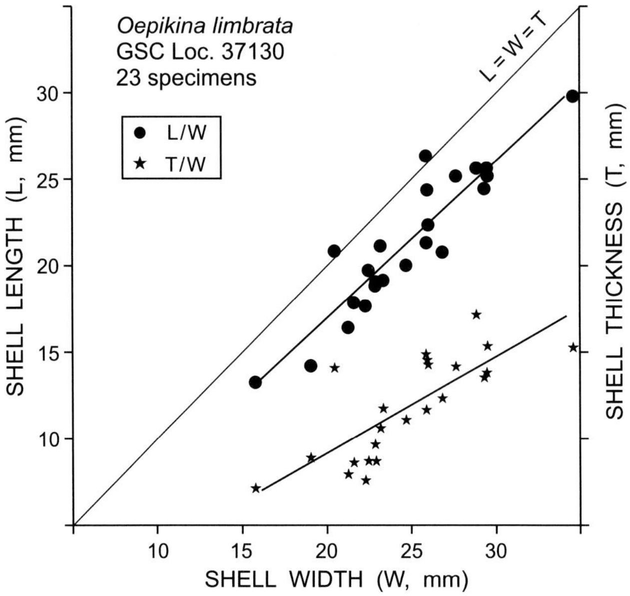

Dorsal interior. Cardinal process bilobed, sitting on posterior part of notothyrial platform, with two discrete lobes projecting ventrally and slightly posteriorly; crest of each lobe triangular in shape, usually bearing fine crenulations. Sockets deep, each with four or five crenulations extending from floor to inner posterior slopes; inner socket ridges high, thick, extending antero-laterally and then laterally, continuous with weaker outer ridges. Median ridge (myophragm) low, thick, posteriorly continuous with notothyrial platform (Pl. 12, figs. 9, 10), bifurcating anteriorly before reaching anterior margin of muscle field, there replaced by much thinner median septa extending near peripheral rim (Pl. 12, figs. 9, 10). Adductor muscle field not clearly defined by bounding ridges, slightly elevated around margin of field, with several low, radiating, transmuscle ridges. Pair of side septa high; their posterior portions inside muscle field stronger than portions anterior of muscle field, extending anteriorly as far as median septa. Peripheral rim low, thick, continuous with lateral ends of socket ridges, interrupted by radiating grooves (crenulations), especially along its anterior portion.

Remarks. Specimens of *Oepikina limbrata* from southern Manitoba are nearly identical to those from Iowa (Wang 1949), particularly in their semicircular, evenly concavo-convex shell with postero-lateral rugae, large, subcircular ventral and dorsal muscle fields with striated floors, and three distinct transmucle septa. The southern Manitoba shells, however, tend to have better developed peripheral rims in both valves and stronger striations in the ventral and dorsal muscle fields than the Iowa forms. Also, the pear-shaped ventral adductor muscle scars figured in Wang (1949) have not been observed in the Manitoba shells. As these features are typically variable and related to shell size and thickness, they are treated as infraspecific variation. The three long transmuscle septa in *Oepikina limbrata* contrast sharply with the four relatively short transmuscle septa in *O. lata* (Pl. 11, figs. 6, 7).

One of the two specimens of *Oepikina* aff. *O. limbrata* Wang from the lower Stony Mountain Formation of eastern Montana (Ross 1957) seems to be more strongly transverse than the southern Manitoba shells. *Oepikina* cf. *O. pergibbosa* (Foerste 1917) from the Bighorn Formation of northern Wyoming (Macomber 1970) shows all the diagnostic features of *O. limbrata* and are probably assignable to this species.

On the basis of their well-developed crenulations (denticles) on the teeth and in the sockets and hypercline dorsal interarea, Dewing (1999) proposed to assign *Oepikina limbrata* to *Mjoesina* Spjeldnaes, 1957. The largely Caradocian *Mjoesina* has strongly crenulated (or striated) teeth and sockets but lacks the characteristic dorsal transmuscle septa of *Oepikina*. All shells of *O. limbrata* from the Hudson Bay Lowlands (Jin et al. 1997), southern Manitoba, and Iowa have well-developed, *Oepikina*-type transmuscle septa.

Order PENTAMERIDA Schuchert and Cooper, 1931

Superfamily PORAMBONITOIDEA Davidson, 1853

Family PARASTROPHINIDAE Ulrich and Cooper, 1938

Genus *Parastrophinella* Schuchert and Cooper, 1931

Type species. *Pentamerus reversus* Billings, 1857. Prinsta Member, Ellis Bay Formation (Hirnantian), Anticosti Island, Québec. (For more details, see Jin and Copper 1997, 2000.)

Parastrophinella cirrita n. sp.

Pl. 14, figs. 14–18; Pl. 22, figs. 3, 4; Fig. 21

Types. Two shells: holotype GSC 117778, (Pl. 14, figs. 14–18) and paratype, GSC 117796 (serially sectioned; Pl. 15, figs. 3, 4; Fig. 21), both from the Selkirk Member (Maysvillian), Red River Formation, Garson Quarry (GSC loc. C-205935), southern Manitoba.

Etymology. From the Latin adjective, *cirritus*, having fine fibres. The feminine adjective *cirrita* depicts the fibrous shell structure of the new species.

Diagnosis. Small shells of *Parastrophinella* with short, sparse costae near anterior margin. Shell wall composed of outer lamellar layer and inner fibrous layer. Median septum in both valves poorly developed. Spondylium sessile posteriorly. Cruralium narrow, sessile anteriorly. Alate plates long and wide.

Description. Shell small, barely reaching medium size, weakly dorsibiconvex, thickest and widest near mid-length. Ventral umbo low, gently convex with erect to suberect beak. Dorsal umbo strongly arched, slightly higher than ventral umbo (Pl. 14, fig. 16), with beak strongly incurved and projecting toward delthyrium (Fig. 21, 0.8 mm from apex). Ventral sulcus confined to anterior one-third of shell, occupying about three-fifths of shell width at anterior margin, bearing three simple, low, rounded costae. Dorsal fold gentle, well defined anteriorly, marked by four simple, rounded costae toward anterior margin. Shell smooth in posterior two-thirds, with anterior portion of each flank marked by two or three gentle undulations.

Shell measurements (*mm*):

	L	*W*	*T*	*L/W*	*T/W*
Holotype GSC 117778	7.91	8.83	5.53	0.90	0.63
Paratype GSC 117796	10.88	12.70	6.81	0.86	0.54

Microscopic shell structure. Shell wall composed of thin outer lamellar layer and thick inner fibrous layer. Calcite fibres crudely arranged lengthwise, diamond-shaped in cross section (Fig. 21; Pl. 22, fig. 4).

Ventral interior. Teeth relatively small, knobby; spondylium wide, deep, broadly V-shaped, sessile posteriorly, supported by thin median septum increasing in height anteriorly.

Dorsal interior. Cruralium narrow, deep, posteriorly supported by low median septum, anteriorly becoming sessile. Median septum very short and low, extending for about 10% of shell length. Inner plates thick, narrow, longer than outer plates. Alate plates prominent (Fig. 21; Pl. 22, fig. 3), narrow in umbonal area, becoming increasingly wider anteriorly, arranged nearly perpendicular to commissural plane, occupying more than one-third of shell thickness (Fig. 21, 3.0 mm from apex) and extending for nearly one-third of shell length.

Remarks. This is the first record of *Parastrophinella* from the Red River Formation of the Williston Basin, although the shells were mentioned by Jin et al. (1997) as *Parastrophinella* sp. in their comparison of the Late Ordovician brachiopod fauna of the Hudson Bay Lowlands with that of the Williston Basin. The well-developed fibrous shell layer is also noted here for the first time for *Parastrophinella*. The small size of this

Fig. 21. Transverse serial sections of *Parastrophinella cirrita* n. sp., paratype, GSC 117796 (mature shell: L = 10.9 mm, W = 12.7 mm, T = 6.81 mm), Garson Quarry (GSC loc. C-205935), Selkirk Member, Red River Formation.

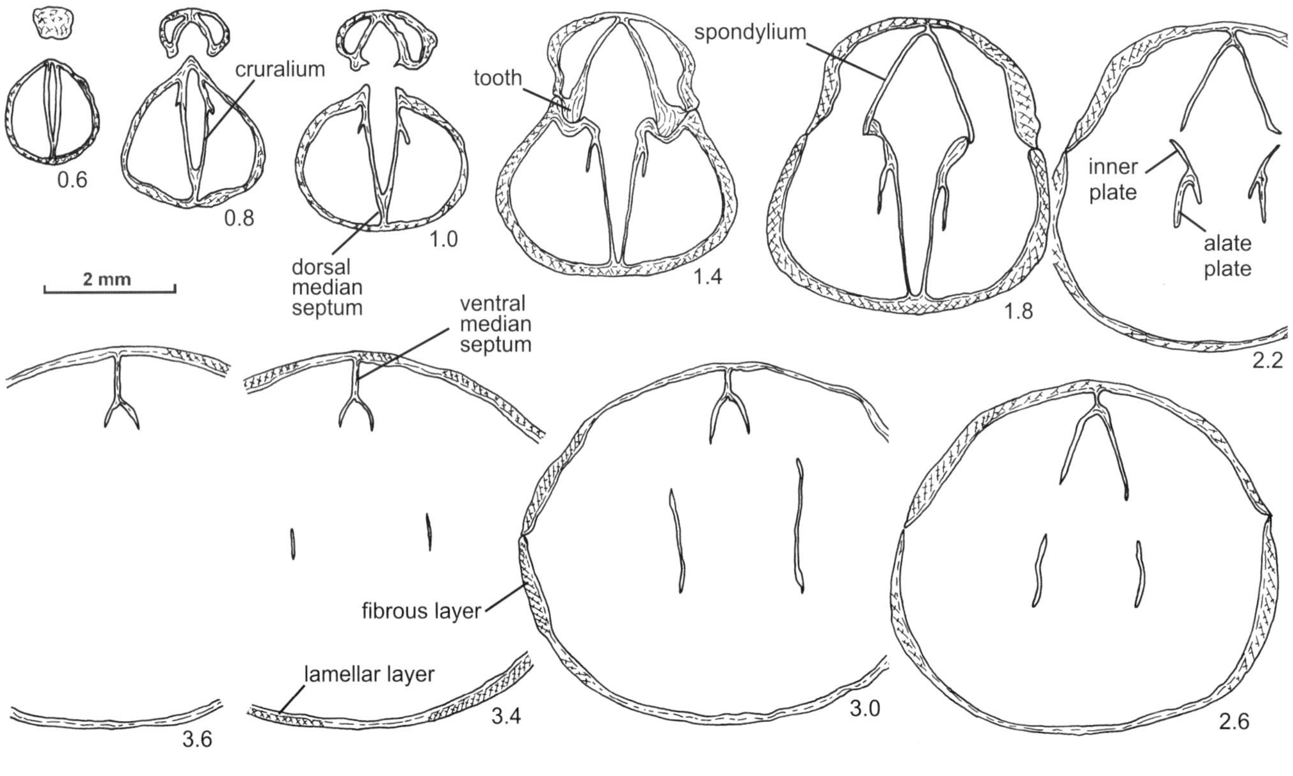

species is distinct among the generally large-shelled Red River brachiopod fauna.

The new species shows superficial similarity to *Parastrophina hemiplicata* (Hall, 1847) in its small shell size and anteriorly developed costae, but its poorly developed median septum in both the ventral and dorsal valves, posteriorly sessile spondylium, and anteriorly sessile cruralium indicate its affinity to *Parastrophinella* rather than to *Parastrophina*. The small shell of the new species resembles the immature forms of *Parastrophinella reversa* from the Ellis Bay Formation of Anticosti Island (Jin and Copper 1997, 2000) in its subcircular outline and anteriorly confined simple, rounded costae. The southern Manitoba shells, however, show a strongly arched dorsal umbo that is slightly higher than the ventral umbo. As in the type species, this "reversed" valve size generally indicates a mature growth stage of the shell. *Parastrophinella portentosa* Nikitin, Popov, and Holmer, 1996, from the Upper Ordovician of central Kazakhstan also has a relatively small shell, but it differs from the new species in its pointed posterior, unusually large, broad, triangular ventral sulcus, and costae confined to the anterior fourth or fifth of the shell. The usually large alate plates and well-developed fibrous shell layer have not been observed in other species of *Parastrophinella*.

Order RHYNCHONELLIDA Kuhn, 1949

Superfamily RHYNCHONELLOIDEA Gray, 1848

Family RHYNCHOTREMATIDAE Schuchert, 1913

Genus *Rhynchotrema* Hall, 1860

Type species. *Atrypa increbescens* Hall, 1847. Trenton Limestone, late Caradoc, New York.

Rhynchotrema increbescens (Hall, 1847)

Pl. 14, figs. 19–22

1847 *Atrypa increbescens* Hall, p. 146, pl. 33, figs. 13a–d.

1860 *Rhynchonella* (*Rhynchotrema*) *increbescens* Hall; Hall, p. 66.

1894 *Rhynchotrema increbescens* (Hall); Hall and Clarke, p. 182 (no illustration, only a revision of description).

1889 *Rhynchonella increbescens* Hall; Nettelroth, p. 83, pl. 34, figs. 26–29.

1893 *Rhynchotrema inaequivalvis* (Castelnau); Winchell and Schuchert, p. 459, pl. 34, figs. 12–14, 20–23, 24–25?.

1923 *Rhynchotrema wisconsinense* Fenton and Fenton, p. 71, pl. 71, figs. 6–8.

1949 *Rhynchotrema increbescens* (Hall); Wang, p. 11 (lectotype selected).

1955 *Rhynchotrema increbescens increbescens* (Hall); Weiss, p. 772, pl. 70, figs. 3–7.

1956 *Rhynchotrema increbescens* (Hall); Cooper, p. 628 (four original type specimens reexamined).

1965 *Rhynchotrema increbescens* (Hall); Schmidt and McLaren, p. 555, fig. 422-3.

1973 *Rhynchotrema increbescens?* (Hall); Alberstadt, p. 51, pl. 7, figs. 6a–e, 7a–e.

1989 *Rhynchotrema increbescens* (Hall); Jin, p. 44, pl. 3, figs. 23–27.

1992 *Rhynchotrema increbescens* (Hall); Jin and Lenz, p. 138, pl. 1, figs. 1–25.

Material. Basal Penitentiary Member (MMMN EPD 970605): 1 complete internal mould.

Remarks. The only specimen in the southern Manitoba collection is a complete internal mould of a conjoined, subtriangular shell, measuring 8.4 mm long, 9.2 mm wide, and 4.1 mm thick. Compared to *Rhynchotrema iowense*, the shell is less strongly transverse and less convex, with a notably narrower ventral sulcus and well-developed septalium. The costae are more numerous, with seven or eight on each shell flank. These features are typical of *Rhynchotrema increbescens.*

Rhynchotrema iowense Wang, 1949

Pl. 15, figs. 1–19; Fig. 22

1949 *Rhynchotrema iowense* Wang, p. 12, pl. 4C, figs. 1–9.

1967 *Rhynchotrema iowense* Wang; Howe, p. 858, pl. 105, figs. 5, 7–9, 11.

1970 *Rhynchotrema iowense* Wang; Macomber, p. 444, pl. 80, figs. 33–47.

1992 *Rhynchotrema iowense* Wang; Jin and Lenz, p. 142, pl. 3, figs. 13–22.

Types: The holotype and paratypes designated by Wang (1949) are from the Brainard Shale, Richmondian (Ashgillian), Iowa.

Southern Manitoba material. Penitentiary Member: 2 complete internal moulds (MMMN I-3091 and MMMN I-3092).

Selkirk Member, Garson Quarry (GSC loc. C-205935): 6 conjoined shells.

Description (southern Manitoba material). Shell small- to medium-sized, transversely elliptical, attaining maximum length of 13.6 mm, width 16.5 mm, and thickness 12.1 mm, with average length/width ratio of 0.8. Lateral profile dorsibiconvex; with maximum thickness and width located at two-thirds of shell length from apex. Ventral umbo evenly convex, non-carinate, pointed, tapering at 90–100°, with suberect to weakly incurved beak. Dorsal umbo more strongly convex than ventral umbo, with beak strongly incurved into delthyrium (Fig. 22, 0.5 mm from apex). Ventral sulcus broad, beginning at 2–4 mm from apex, widening rapidly anteriorly to occupy about one-half shell width at anterior margin. Dorsal fold broad, low, with flat top. Entire shell covered by strong, simple, subangular costae, with three in sulcus, four on fold, five or six on each flank. Concentric growth lines well developed, evenly spaced over entire shell surface, becoming strong, zigzag lines near anterior margin of relatively large shells (Pl. 15, figs. 14, 19).

Shell measurements (mm):

	L	*W*	*T*	*L/W*	*T/W*
GSC 117779	12.8	16.6	12.1	0.77	0.73
GSC 117780	9.5	12.2	7.4	0.78	0.61

Ventral interior. Teeth large, robust (Fig. 22); dental plates high, subparallel to each other, largely confined to umbonal cavity. Muscle field clearly defined, posteriorly bounded by low, ridge-like extensions of dental plates, open anteriorly.

Dorsal interior. Cardinal process low, short, blade-like (Fig. 22); septalium small, shallow; hinge plates stout, bounding shallow sockets. Median septum strong posteriorly, becoming low ridge anteriorly, extending for slightly less than half shell length. Crura rod-like. Adductor muscle scars poorly impressed.

Remarks. Specimens from southern Manitoba are assigned to *Rhynchotrema iowense* on the basis of their transversely elliptical shell, strong and simple costae, and broad fold with a flat top. Some very small shells are nearly equidimensional (with a length/width ratio of 0.93), but adult shells show accelerated widening with ontogeny. The southern Manitoba shells are virtually identical to those of the same species from Iowa (Wang 1949) and Wyoming (Macomber 1970) in both

Fig. 22. Transverse serial sections of *Rhynchotrema iowense* Wang, 1949, hypotype, GSC 117801 (adult shell: L = 14.1 mm, W = 16.5 mm, T = 9.8 mm), Garson Quarry, Selkirk Member, Red River Formation.

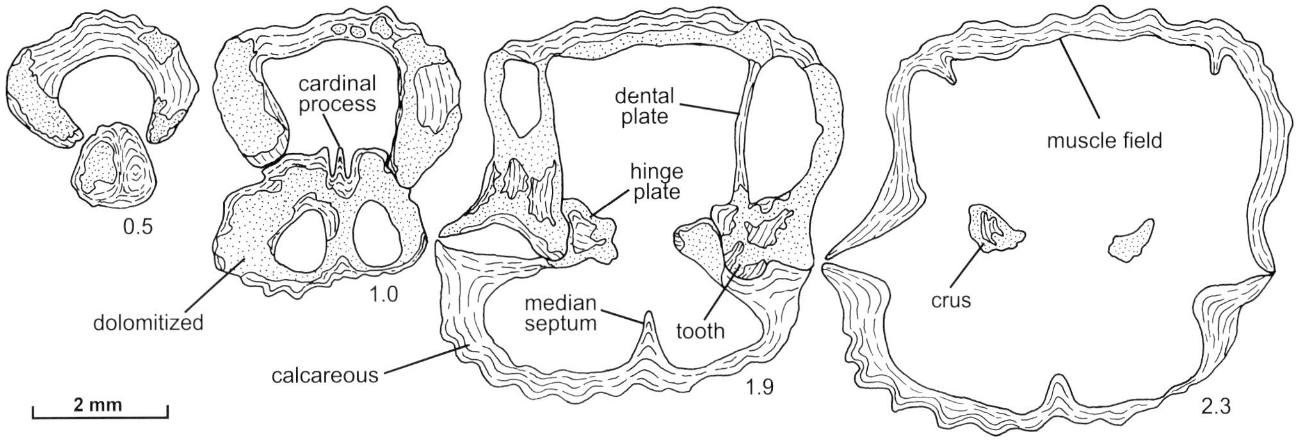

external morphology and internal characters. The shells reported as *R. iowense* by Ross (1957) from the uppermost Bighorn Dolomite of Wyoming have a much weaker fold and sulcus than the typical forms. Jin (1989) noted that most species of *Rhynchotrema* have well-developed septalial plates forming a relatively large and deep septalium. In the serially sectioned specimen of *R. iowense* from southern Manitoba, however, the septalial plates do not show up clearly, although this appears to have been the result of obliteration by dolomitization of the shell (Fig. 22). The only two specimens from the Stony Mountain Formation are notably smaller than those from the Red River Formation.

Genus *Lepidocyclus* Wang, 1949

Type species. *Lepidocyclus laddi* Wang, 1949. Upper Elgin Member, Maquoketa Formation, Maysvillian (early Ashgill), Iowa.

Lepidocyclus laddi Wang, 1949

Pl. 16, figs. 1–5

Southern Manitoba material. Selkirk Member, (GSC 117782): 1 conjoined shell.

Remarks. This specimen is assigned to *Lepidocyclus* on the basis of its strongly lamellose shell and well-developed deltidial plates. The shell is nearly identical to those of the type species from Iowa (Wang 1949) in its medium size (20.8 mm long, 22.0 mm wide, and 17.9 mm thick), nearly equal length and width, dorsibiconvexity, and incurved ventral beak. The only difference is that the Manitoba shell is more clearly pentagonal in outline and more strongly dorsibiconvex than the Iowa forms. These features, however, are typically variable at different growth stages and thus regarded as infraspecific variation.

Genus *Hypsiptycha* Wang, 1949

Type species. *Hypsiptycha hybrida* Wang, 1949. Brainard Member (Richmondian), Maquoketa Formation, northwestern Maquoketa, Iowa.

Remarks. Both *Hypsiptycha* and *Lepidocyclus* have well-developed deltidial plates, which separate them from *Hiscobeccus*. *Hypsiptycha* can be distinguished from *Lepidocyclus* in that the shell of *Hypsiptycha* tends to be elongate oval, subpentagonal to subtriangular and, more importantly, its septalium is nearly obsolete, with the septalial plates raised ventrally into a nearly horizontal plate (similar to a cardinal plate) at the level of the hingeline (Pl. 22, figs. 5, 6; Pl. 23, figs. 1, 4, 5), thus displacing the commonly inflated cardinal process well into the ventral valve (Jin 1989; Jin et al. 1989; Jin and Lenz 1992). *Lepidocyclus* typically has a globose shell, much deeper septalium and a thinner, blade-like cardinal process.

Hypsiptycha anticostiensis (Billings, 1862)

Pl. 15, figs. 20–24; Pl. 25, figs. 6–17; Pl. 22, figs. 5, 6; Pl. 23, fig. 1; Fig. 23

1862 *Rhynchonella Anticostiensis* [sic] Billings, p. 142, figs. 119a–c.

1928 *Rhynchotrema anticostiense* (Billings); Twenhofel, p. 207, pl. 21, figs. 4–6.

1943 *Rhynchotrema anticostiense* (Billings); Okulitch, p. 74, pl. 1, figs. 5–6.

1957 *Hypsiptycha* cf. *H. anticostiense* (Billings); Ross, p. 478, pl. 39, figs. 6–8, 12–13.

1967 *Hypsiptycha* cf. *H. anticostiensis* (Billings); Howe and Reso, p. 358, pl. 40, figs. 17–20.

1970 *Hypsiptycha anticostiensis* (Billings); Macomber, p. 445, pl. 80, figs. 18–22, 27–32.

1977 *Hypsiptycha* cf. *H. anticostiensis* (Billings); Mitchell, p. 120, pl. 26, figs. 26–33.

1989 *Hypsiptycha anticostiensis* (Billings); Jin, p. 68, pl. 9, figs. 1–12; pl. 26, figs. 6–10.

1997 *Hypsiptycha anticostiensis* (Billings); Jin et al., p. 37, pl. 24, figs. 1–17; pl. 25, figs. 1–9.

Types. The lectotype, GSC 2032d, selected by Howe and Reso (1967), came from the Vauréal Formation (Richmondian), Anticosti Island.

Southern Manitoba material. Selkirk Member, Garson Quarry (GSC loc. C-205935): 4 conjoined shells.

Description (southern Manitoba material). Shell medium-sized, subtriangular to subpentagonal in outline, with nearly equal length and width. Lateral profile dorsibiconvex, thickest at about mid-length. Ventral umbo narrow, with apical angle of 80–90°, uniformly convex, non-carinate, protruding high above hingeline, and tapering into suberect to erect beak. Deltidial plates thick, weakly convex posteriorly, becoming concave anteriorly to accommodate incurved ventral beak (Fig. 23, 0.6 through 2.1 mm from apex). Dorsal umbo strongly convex, with beak curved into delthyrial cavity. Median groove on dorsal umbo changing into moderately high fold anteriorly. Sulcus moderately deep, occupying about 40% of shell width at anterior margin. Fold and sulcus originating about 3–4 mm anterior of valve apex. Shell costae simple, rounded, strongest on fold and in sulcus, usually four on fold, three in sulcus, and six or seven on each flank. Growth lamellae well developed, unevenly distributed on the entire shell surface, averaging 1–2 per mm in anterior part of shell. Finer concentric growth lines also well developed, much clearer in the groove between costae, averaging 9 per mm.

Shell measurements (*mm*):

	L	*W*	*T*	W_2	*L/W*	*T/W*
GSC 117784	18.1	18.2	12.7	8.7	0.99	0.70
GSC 117781	14.9	14.4	10.8	—	1.03	0.75
GSC 117783	15.8	15.6	11.2	7.1	1.02	0.72
GSC 117797	17.1	15.2	13.2	8.0	1.12	0.86

Fig. 23. Transverse serial sections of *Hypsiptycha anticostiensis* (Billings, 1862), hypotype, GSC 117797 (mature shell: L = 17.5 mm, W = 15.1 mm, T = 13.8 mm), Garson Quarry, Selkirk Member, Red River Formation.

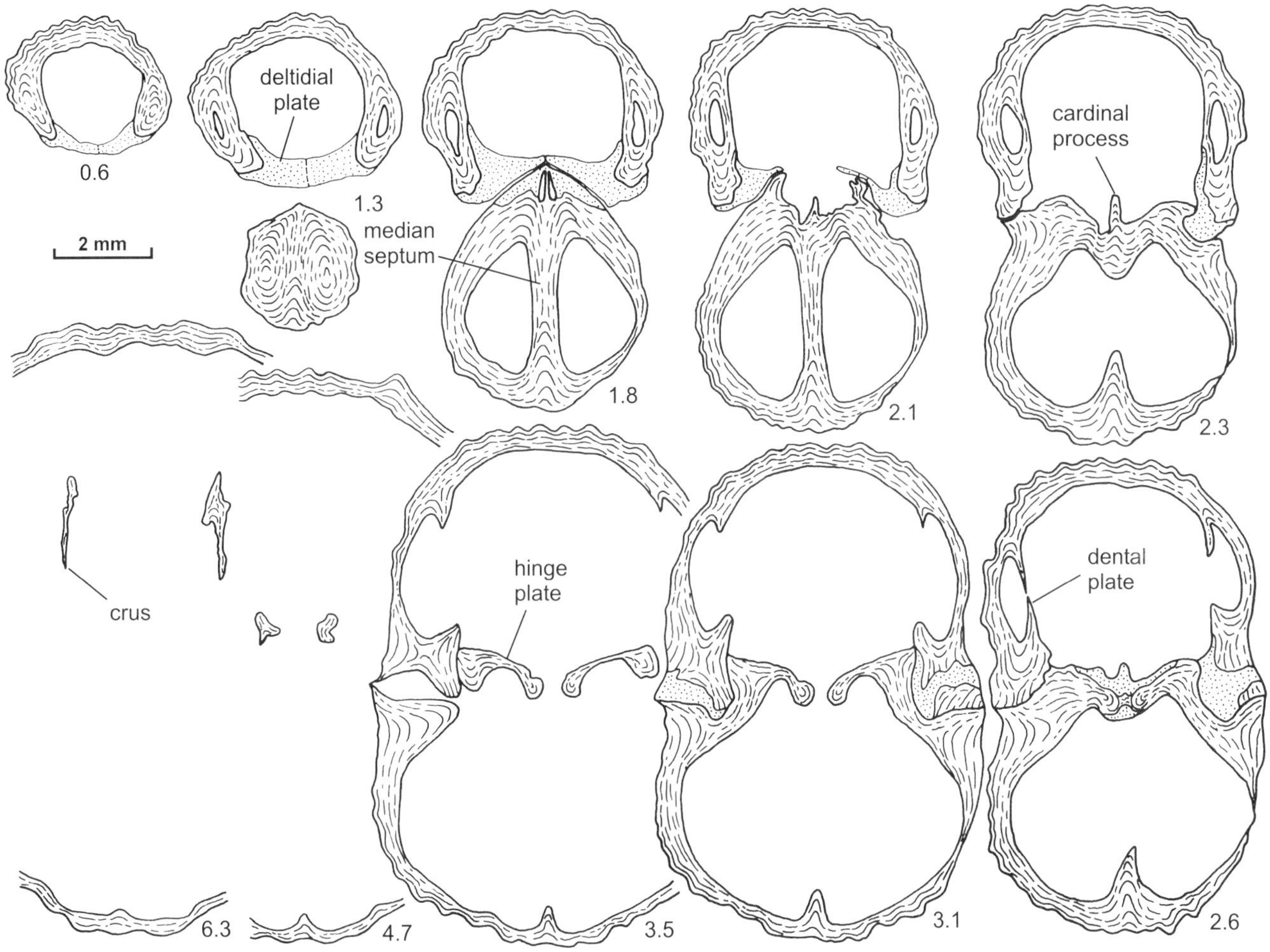

Ventral interior. Teeth large, massive, laterally located (Fig. 23, 2.3 through 3.1 mm from apex), and supported partially by thickened lateral shell wall. Dental plates thin, low, forming small lateral cavities. Muscle field bounded posteriorly by extensions of dental plates, anteriorly becoming slightly elevated above valve floor.

Dorsal interior. Sockets large, broad, with prominent inner socket ridges. Cardinal process blade-like, slightly thickened at anterior end. Septalium very shallow to obsolete, with septalial plates almost parallel to commissural plane and raised to level of hingeline. Median septum high in umbonal area, extending anteriorly as low median ridge of adductor muscle field (Fig. 23, 3.5 through 6.3 mm from apex). Hinge plates well developed, almost horizontal. Crura radulifer, extending for slightly more than half shell length, strongly curved ventrally. Adductor scars clearly impressed, wider anteriorly, lacking bounding ridges.

Remarks. Hypsiptycha anticostiensis has a wide distribution in Upper Ordovician rocks of the Laurentia paleoplate, being known from Anticosti Island, Hudson Bay Lowlands, southern Manitoba, eastern Montana (Ross 1957), southwestern Nevada (Howe and Reso 1967), Wyoming (Macomber 1970), and northern Ireland (Mitchell 1977).

Hypsiptycha occidens (Wilson, 1926)

Pl. 17, figs. 1–19; Pl. 18, figs. 1–7; Pl. 23, figs. 4, 5; Figs. 24–26

1926 *Rhynchotrema increbescens* var. *occidens* Wilson, p. 21, pl. 4, figs. 6–10.

1926 *Rhynchotrema pisina* Wilson, p. 22, pl. 4, fig. 13 (*non* figs. 11, 12).

1989 *Hypsiptycha occidens* (Wilson); Jin et al., p. 26, pl. 2.1, figs. 5–26; pl. 2.2, figs. 1–10; pl. 2.9, figs. 1–8.

1992 *Hypsiptycha occidens* (Wilson); Jin and Lenz, p. 105, pl. 7, figs. 19–21; pl. 8, figs. 1–12.

1997 *Hypsiptycha occidens* (Wilson); Jin et al., p. 38, pl. 23, figs. 1–24.

Types. The lectotype (GSC 6746), selected by Jin et al. (1989), is from the lower Beaverfoot Formation (Ashgill), Rocky Mountains, British Columbia.

Southern Manitoba material. Penitentiary Member (GSC loc. O-27186 and 55-60): 2 dorsal internal, 2 ventral internal, and 2 complete internal moulds.

Gunn Member (GSC loc. O-27185, O-37130, C-205926, C-205928, and C-205930): 366 conjoined shells, 11 dorsal and 8 ventral valves.

Selkirk Member (MMMN I-2680, I-2788): 1 conjoined shell, 1 complete internal mould, 1 dorsal external and 1 ventral internal moulds.

Description (southern Manitoba material). Shell small to medium-sized, with average length 9.4 mm (maximum 16.3 mm), width 9.4 mm (maximum 14.8 mm), thickness 6.2 mm (maximum 11.4 mm), and length/width ratio close to 1 (Figs. 24, 25). Lateral profile biconvex to strongly dorsibiconvex, attaining greatest thickness at about two-thirds of shell length from apex. Ventral sulcus and dorsal fold well developed, beginning at about one-third of shell length from apex, and width of sulcus occupying one-half of shell width at anterior margin. Ventral umbo pointed, tapering at apical angles of 80–90°, terminating into erect to suberect beak. Deltidial plates thick, strongly convex posteriorly, becoming concave at anterior end (Fig. 26, 0.5–1.9 mm from apex); foramen submesothyridid. Costae strong, rounded, simple, three in sulcus, four on fold, and about six to eight on each shell flank. Concentric growth lamellae well developed, strongest in anterior part of shell, averaging 3 per 1 mm near anterior margin.

Statistics of shell measurements (*mm*):

GSC loc. O-37130	*L*	*W*	*T*	W_1	*L/W*	W_1/W
AVG	9.4	9.4	6.2	4.7	1.01	0.50
STD	1.8	1.8	1.7	1.2	0.08	0.05
MIN	3.3	6.0	2.8	2.1	0.45	0.35
MAX	16.3	14.8	11.4	9.8	1.17	0.66

Fig. 24. Shell dimensions of *Hypsiptycha occidens* (Wilson, 1926), sample from GSC loc. O-37130, Gunn Member, Stony Mountain Formation. Note that *T/W* ratios shell convexity) are more variable than *L/W* ratios.

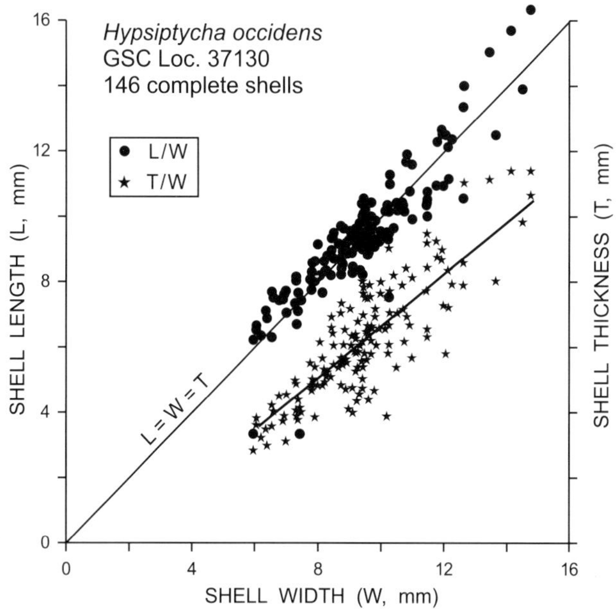

Ventral interior. Teeth strong, massive; dental plates relatively thin and short, partly fused to lateral shell wall to form minute dental cavities (Fig. 26, 1.7–2.5 mm from apex). Muscle field poorly impressed, marked posteriorly by low extensions of dental plates.

Dorsal interior. Sockets large, deep, with large and high inner socket ridges. Cardinal process plate-like; septalium obsolete, with septalial plates transformed

Fig. 25. Shell dimensions of *Hypsiptycha occidens* (Wilson, 1926), sample from GSC loc. O-37130, Gunn Member, Stony Mountain Formation. Note that the sulcus width/shell width (W_2/W) ratio is relatively consistent during ontogeny.

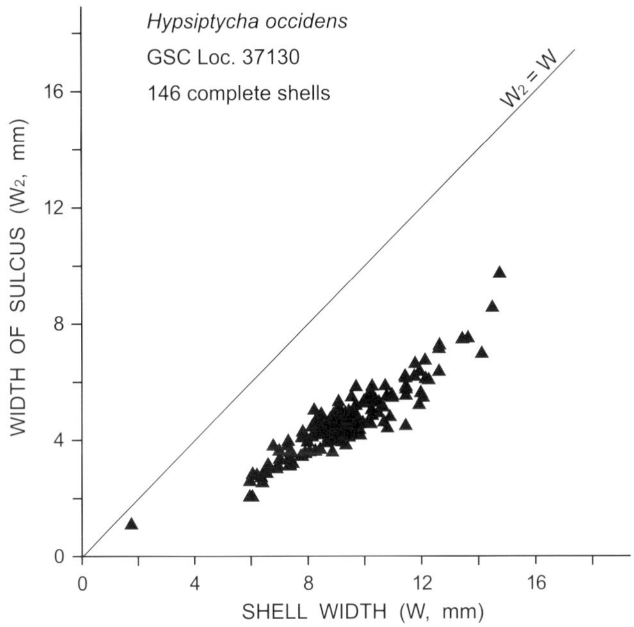

into ventrally arched to horizontal plate (resembling cardinal plate) to support cardinal process. Median septum thick and high in notothyrial cavity, becoming thinner anteriorly, and continuing anteriorly to become myophragm of adductor muscle field. Crura strong, rod-like, radulifer.

Remarks. Examination of hundreds of specimens in the southern Manitoba collection reveals a number of infraspecific and ontogenetic variations: (1) the shell convexity (measured by T/W ratio) has a wider range of variation than shell width among shells of similar length (Figs. 22, 23), although smaller individuals tend to have a lower T/W ratio; (2) the number of costae associated with the fold and sulcus is fairly consistent — among 319 individuals from GSC loc. O-37130, three shells have five costae on the fold (four in sulcus), one shell has three costae on the fold (two in sulcus), and all the others show four on the fold (three in sulcus); and (3) the dental plates in small shells are typically thin, free from lateral shell walls, but in larger shells, they are nearly fused to the shell walls, leaving only minute dental cavities.

Fig. 26. Transverse serial sections of *Hypsiptycha occidens* (Wilson, 1926), hypotype, GSC 117799 (mature shell: L = 15.2 mm, W = 13.2 mm, T = 11.1 mm), GSC loc. O-37130, Gunn Member, Stony Mountain Formation.

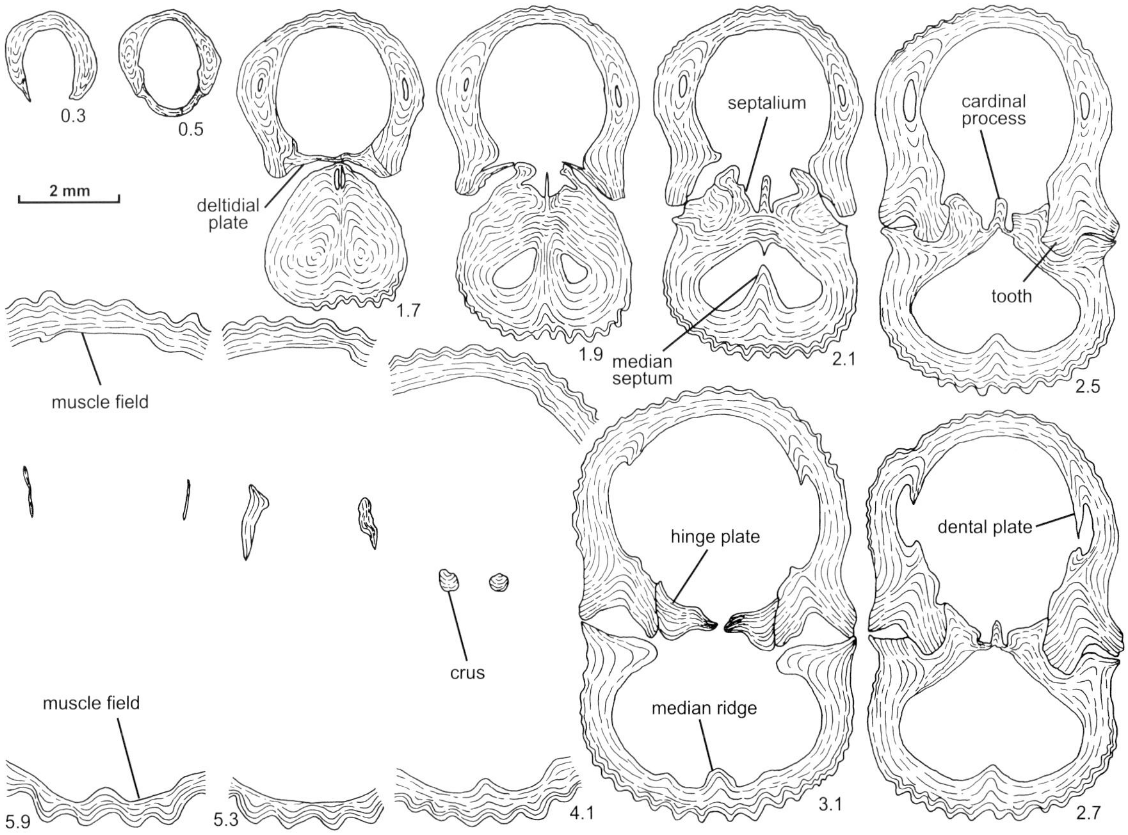

Hypsiptycha occidens is distinguished from the type species by its notably smaller shell, which tends to be subtriangular with a more pointed posterior. *Hypsiptycha* cf. *H. neenah* (Macomber 1970) from the Bighorn Formation of Wyoming differs from *H. occidens* in its unusually strong fold, elevated median costae, and sharp costae.

Genus *Hiscobeccus* Amsden, 1983

Type species. *Atrypa capax* Conrad, 1842. Richmondian rocks, Richmond, Indiana. Precise locality and stratum for the neotype designated by Amsden (1983) have not been determined.

Hiscobeccus capax (Conrad, 1842)

Pl. 18, figs. 8–17; Pl. 23, figs. 2, 3; Fig. 27

1842 *Atrypa capax* Conrad, p. 264, pl. 14, fig. 21.

1847 *Atrypa increbescens* Hall, p. 146, pl. 33, figs. 13t–y.

1860 *Rhynchotrema capax* (Conrad); Hall, pp. 66–68.

1873 *Rhynchonella capax* (Conrad); Meek, p. 123, pl. 11, figs. 2a–f.

1880 *Rhynchonella capax* (Conrad); White, p. 489, pl. 1, figs. 8–11.

1882 *Rhynchonella capax* (Conrad); Whitfield, p. 263, pl. 12, figs. 26, 27.

1893 *Rhynchotrema capax* (Conrad); Winchell and Schuchert, p. 462, pl. 34, figs. 30–33.

1894 *Rhynchotrema capax* (Conrad); Hall and Clarke, p. 182, pl. 65, figs. 14–16, 21–23.

1966 *Lepidocyclus capax* (Conrad); Howe, p. 263, pl. 31, figs. 15–20.

1973 *Lepidocyclus capax* (Conrad); Alberstadt, p. 53, pl. 6, figs. 1–3.

1983 *Hiscobeccus capax* (Conrad); Amsden, pp. 38–39, pl. 4, figs. 1a–e; pl. 7, figs. 2a–d.

1997 *Hiscobeccus capax* (Conrad); Jin et al., p. 38, pl. 27, figs. 1–12; pl. 28, figs. 1–12; pl. 29, fig. 1.

Types. The neotype, UC 12403b (Field Museum of Natural History), designated by Amsden (1983), came from Richmondian rocks, Richmond, Indiana.

Southern Manitoba material. Selkirk Member (MMMN I-2767): 6 conjoined shells and 1 ventral valve.

Cat Head Member (MMMN I-2906): 1 conjoined shell.

Description (southern Manitoba material). Shell medium-sized, barely reaching large size, suboval to subtriangular in outline, strongly dorsibiconvex to globose in lateral view, with average length 16.1 mm (maximum 21.2 mm), width 18.3 mm (maximum 22.7 mm), thickness 13.8 mm (maximum 17.4 mm), length/width ratio 0.91, and thickness/width ratio 0.74; shell attaining greatest width and thickness at about two-thirds of shell length from apex. Ventral umbo narrow with incurved beak. Dorsal umbo broad, strongly arched, with beak curved into delthyrium. Deltidial plates absent. Fold and sulcus well developed, beginning immediately anterior of umbonal areas, occupying about 45% of shell width at anterior margin. Shell costae strong, simple, subrounded, typically four on fold, three in sulcus, and six to eight on each flank. Growth lamellae well developed, most conspicuous in anterior part of shell, averaging 3–4 per 1 mm.

Ventral interior. Teeth small, wedge-like, supported by thickened shell wall. Dental plates absent. Muscle field deeply impressed.

Dorsal interior. Hinge sockets deep. Cardinal process short, blade-like (Fig. 27). Septalium small, shallow. Median septum high and thick posteriorly, extending anteriorly to become low myophragm. Crura radulifer, extending antero-ventrally. Adductor scars well impressed, without bounding ridges.

Remarks. The internal structures of *Hiscobeccus capax* are nearly identical to those of *H. gigas*. Externally, however, *H. gigas* has a notably lower convexity, and much stronger and coarser growth lamellae. The specimens of *H. capax* from the Red River Formation are generally smaller than the type specimen (Conrad 1842) and those from the Hudson Bay Lowlands (Jin et al. 1997). The cardinal process in the sectioned southern Manitoba shell appears to be weaker (also lacking an inflated crest) than that of *Hiscobeccus* from other regions (e.g., Jin 1989; Jin and Lenz 1992). The shells of *Hiscobeccus capax* from the uppermost Bighorn Dolomite of Wyoming (Ross 1957) are identical to the Red River material except for their larger shell size.

Fig. 27. Transverse serial sections of *Hiscobeccus capax* (Conrad, 1842), hypotype, GSC 117798 (mature shell: L = 14.5 mm, W = 15.4 mm, T = 11.3 mm), Garson Quarry, Selkirk Member, Red River Formation.

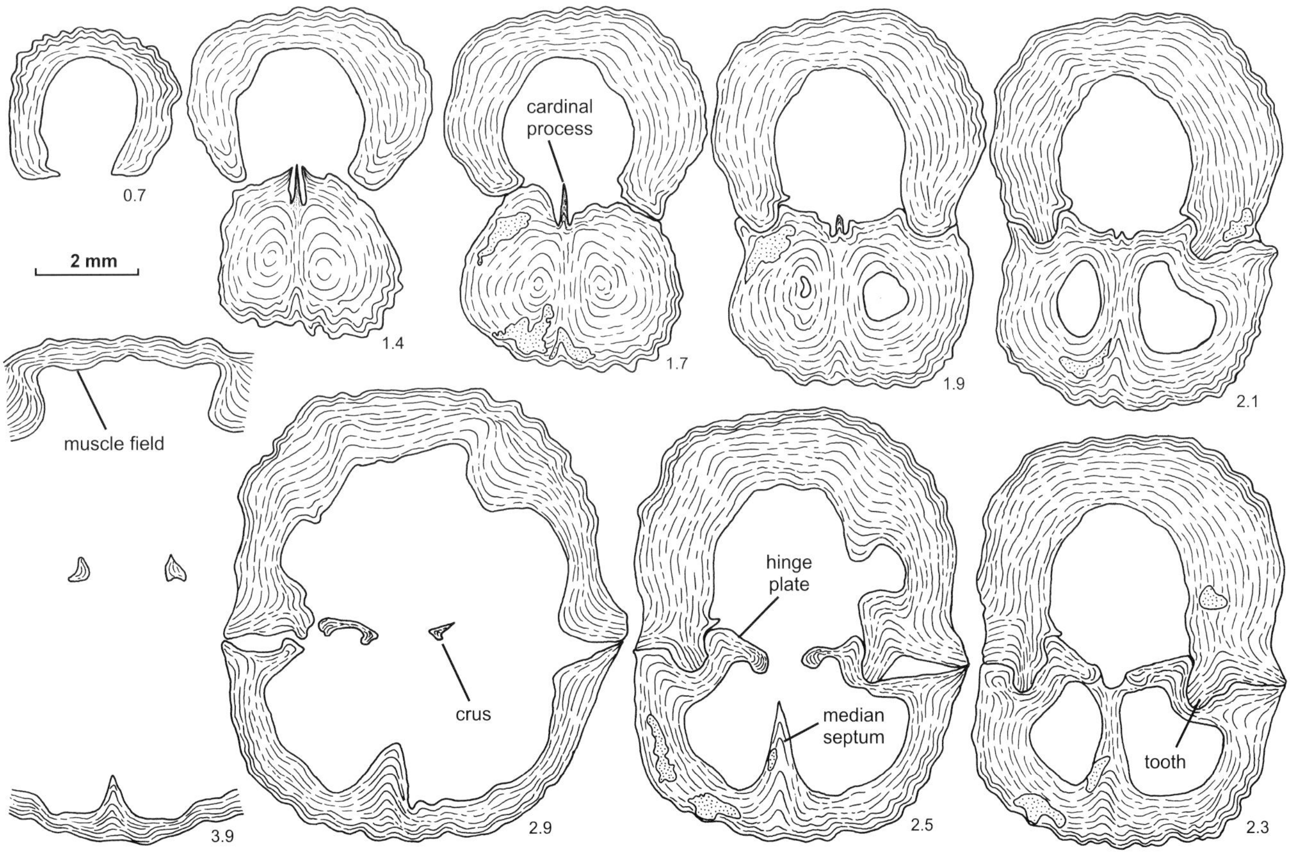

Hiscobeccus gigas (Wang, 1949)

Pl. 19, figs. 1–11; Pl. 20, figs. 8–13; Pl. 21, fig. 1; Figs. 28–30

1949 *Lepidocyclus gigas* Wang, p. 16, pl. 10D, figs. 1–5.

1957 *Lepidocyclus perlamellosa* (part); Ross (*non* Whitfield, 1878), p. 477, pl. 39, figs. 1–5.

1957 *Lepidocyclus capax*; Ross (non Conrad, 1842), p. 477, pl. 39, figs. 21, 24–27.

1970 *Lepidocyclus gigas* Wang; Macomber, p. 447, pl. 80, figs. 1–17.

1981 *Lepidocyclus capax*; Bolton (non Conrad, 1842), pl. 3, fig. 1.

1989 *Lepidocyclus gigas* Wang; Jin, p. 65, pl. 9, figs. 13–22; pl. 10, figs. 1–5; pl. 26, figs. 4, 5.

1997 *Hiscobeccus gigas* (Wang); Jin et al., p. 39, pl. 29, figs. 2–14.

Types. Wang's type lot is from the Brainard Member (Richmondian), Maquoketa Formation, Iowa.

Southern Manitoba material. Penitentiary Member (GSC loc. O-27186 and 55-60): 4 dorsal internal and 8 external, 10 ventral internal and 7 external moulds, and 25 complete internal mounds.

Gunn Member (GSC loc. O-37130, C-205926, C-205928, C-205930, and C-205931): 66 conjoined shells, 1 dorsal and 16 ventral valves.

Description (southern Manitoba material). Shell usually large, transversely subpentagonal to subelliptical, dorsibiconvex with moderate convexity (average thickness/width ratio 0.6), with average length of 24.8 mm, width 27.5 mm, and thickness 16.8 mm (Figs. 28, 29). Ventral umbo low, with strongly incurved beak appressed onto dorsal umbo; delthyrium open; foramen mesothyridid. Dorsal umbo more strongly convex than ventral umbo, with beak curved into delthyrial cavity. Ventral sulcus and dorsal fold beginning slightly anterior of umbonal areas, becoming wider and more pronounced anteriorly, forming short but conspicuous tongue at anterior margin. Costae simple, subrounded, four (rarely three or five) on fold, three (rarely two or four) in sulcus, and six to eight (commonly seven) on each flank. Concentric growth lamellae strong, imbricating, covering entire shell surface but strongest in anterior half of shell.

Statistics of shell measurements (*mm*):

GSC loc. O-37130	*L*	*W*	*T*	W_2	*L/W*	*T/W*	W_2/W
AVG	24.8	27.5	16.76	13.4	0.91	0.60	0.49
STD	5.2	5.7	4.83	3.3	0.07	0.09	0.06
MIN	11.4	12.3	6.40	5.5	0.76	0.45	0.39
MAX	31.9	34.1	26.81	20.7	1.07	0.81	0.65

Ventral interior. Teeth strong, knobby, supported directly by thickened lateral shell wall (Fig. 30, 3.6 mm from apex). Dental plates absent or fused with lateral shell walls. Muscle field deeply impressed, flabellate, defined by steep lateral slopes, with striated floor (Pl. 20, figs. 8, 10, 12); adductor muscle scars elongate oval, located in postero-medial portion of muscle field, completely enclosed by much larger diductor scars.

Fig. 28. Shell dimensions of *Hiscobeccus gigas* (Wang, 1949), sample from GSC loc. O-37130), Gunn Member, Stony Mountain Formation.

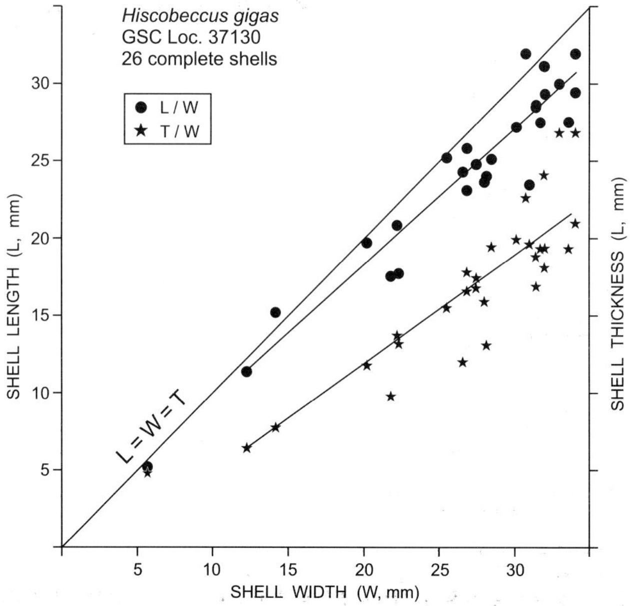

Fig. 29. Shell dimensions of *Hiscobeccus gigas* (Wang, 1949). The same specimens as those for Fig. 25. Note consistent average W_2/W ratio throughout ontogeny.

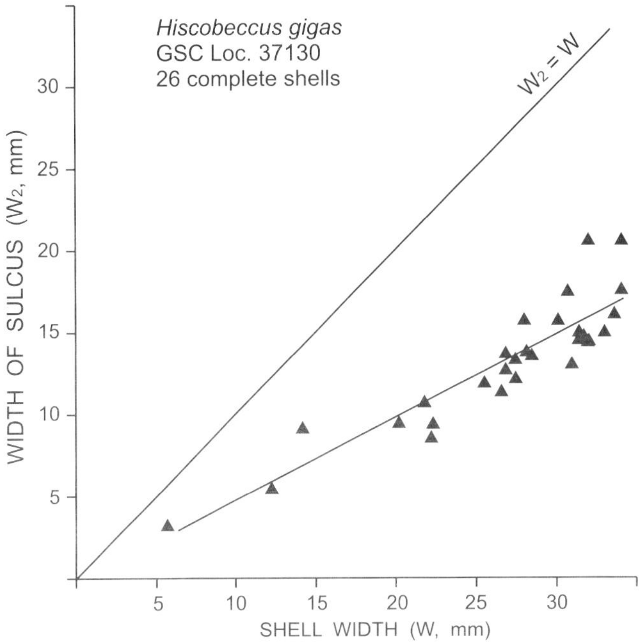

Dorsal interior. Hinge sockets broad, deep, with strong and thick inner socket ridges. Cardinal process high, blade-like (Pl. 20, figs. 9, 13). Septalium small, shallow; median septum short, thick, and high posteriorly, becoming lower anteriorly and continuous with myophragm of muscle field. Adductor muscle field elongate oval, bounded by low, inconspicuous ridges, occupying slightly less than one-half of valve length (Pl. 20, figs. 9, 11, 13). Crura slender, rod-like, radulifer.

Remarks. *Hiscobeccus gigas* attains the largest shell size among Ordovician rhynchonellids, with largest shells from the Stony Mountain Formation of southern Manitoba and the Bighorn Formation of Wyoming reaching more than 40 mm in length or width. Despite the large size, *H. gigas* has a considerably lower biconvexity in comparison to other relatively large-shelled species of *Hiscobeccus*. *Hiscobeccus capax* from the American mid-continental basin (Amsden 1983) and the Hudson Bay Lowlands (Jin et al. 1997), for example, has a globose shell that may be thicker than long in

Fig. 30. Transverse serial sections of *Hiscobeccus gigas* (Wang, 1949), hypotype, GSC 117792 (mature shell: L = 24.4 mm, W = 28.2 mm, T = 19.9 mm), GSC loc. O-37130, Gunn Member, Stony Mountain Formation.

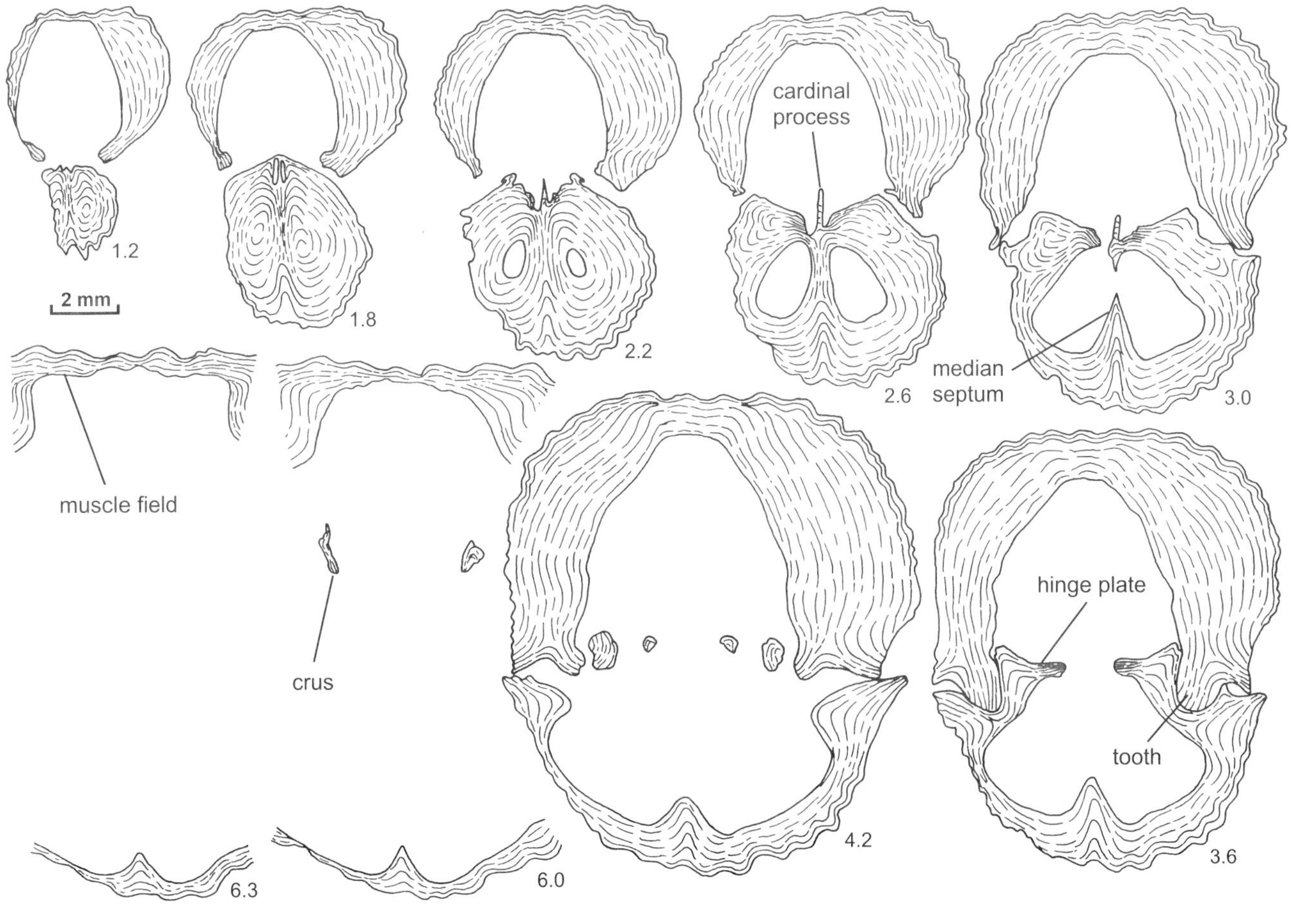

gerontic forms. Compared to the type specimen of *Hiscobeccus gigas* from the Brainard Member of the Maquoketa Formation of Iowa, the southern Manitoba specimens have a greater maximum size. Shells of *H. gigas* from the Bighorn Formation of Wyoming (Macomber 1970) and the Vauréal Formation of Anticosti Island (Jin 1989) are essentially identical to those in the present collection.

References

Alberstadt, L.P. 1973. Articulate brachiopods of the Viola Formation (Ordovician) in the Arbuckle Mountains, Oklahoma. Oklahoma Geological Survey, Bulletin 117, 90 pp.

Amsden, T.W. 1974. Late Ordovician and Early Silurian articulate brachiopods from Oklahoma, southwestern Illinois, and eastern Missouri. Oklahoma Geological Survey, Bulletin 119, 154 pp.

Amsden, T.W. 1983. Upper Bromide Formation and Viola Group (Middle and Upper Ordovician) in eastern Oklahoma. Pt. 3, the Late Ordovician brachiopod genera *Lepidocyclus* and *Hiscobeccus*. Oklahoma Geological Survey, Bulletin 132: 36–44.

Baillie, A.D. 1952. Ordovician geology of Lake Winnipeg and adjacent area, Manitoba. Manitoba Department of Mines and Natural Resources, Mines Branch Publication 15-6, 64 pp.

Bancroft, B.B. 1929. Some new genera and species of Strophomenacea from the Upper Ordovician of Shropshire. Memoirs and Proceedings of the Manchester Literary and Philosophical Society No. 73, pp. 33–65.

Billings, E. 1857. Report for year 1856. Geological Survey of Canada, Report of Progress for the Years 1853–54–55–56, pp. 247–345.

Billings, E. 1860. Description of some new species of fossils from the Lower and Middle Silurian rocks of Canada. The Canadian Naturalist and Geologist, **5**: 49–69.

Billings, E. 1862. New species of fossils from different parts of the Lower, Middle, and Upper Silurian rocks of Canada. Geological Survey of Canada, Palaeozoic Fossils, **1**(4): 96–168.

Bolton, T.E. 1972. Geological map and notes on the Ordovician and Silurian litho- and biostratigraphy, Anticosti Island, Quebec. Geological Survey of Canada, Paper 71-19, 45 pp.

Bolton, T.E. 1977. Ordovician megafauna, Melville Peninsula, southeastern District of Franklin. Geological Survey of Canada, Bulletin 269: 23–75.

Bolton, T.E. 1981. Ordovician and Silurian biostratigraphy, Anticosti Island, Québec. *In* IUGS Field Meeting, Anticosti–Gaspé, Québec, Vol. 2, Stratigraphy and Paleontology. *Edited by* P.J. Lespérance. Université de Montréal, Montréal, pp. 41–59.

Boucot, A.J. 1975. Evolution and extinction rate controls. Elsevier, Amsterdam.

Boucot, A.J. 1983. Does evolution take place in an ecological vacuum? II. Journal of Paleontology, **57**: 1–30.

Boucot, A.J., and Lawson, J.D. (*Editors*). 1999. Paleocommunities: a case study from Silurian and Lower Devonian. Cambridge University Press, Cambridge.

Brett, C.E., Boucot, A.J., and Jones, B. 1993. Absolute depths of Silurian benthic assemblages. Lethaia, **26**: 25–40.

Brindle, J.E. 1960. The faunas of the Lower Palaeozoic carbonate rocks in the subsurface of Saskatchewan. Saskatchewan Department of Mineral Resources, Geology Division, Report 52, 45 pp.

Butler, R.J., Battin, R.L., Plank, R.F., and Winston, G.O. 1955. Lithographic correlation of middle and lower Paleozoic rocks. North Dakota Geological Society Guidebook, South Dakota Black Hills, 3rd Field Conference, pp. 38–42.

Calef, C.E., and Hancock, N.J. 1974. Wenlock and Ludlow marine communities in Wales and the Welsh Borderland. Palaeontology, **17**: 779–810.

Canter, K.L. 1998. Facies, cyclostratigraphic and secondary diagenetic controls on reservoir distribution, Ordovician Red River Formation, Midale Field, southern Saskatchewan. *In* 8th International Williston Basin Symposium, Core Workshop Volume. Saskatchewan, North Dakota & Montana Geological Societies, pp. 41–65.

Carlson, C.G., and Thompson, S.C. 1987. Stratigraphy of the Deadwood Formation and the Winnipeg Group in the Williston Basin. Rocky Mountain Association of Geologists, Denver, pp. 71–81.

Caster, K.E., Dalvé, E.A., and Pope, J.K. 1961. Elementary guide to the fossils and strata of the Ordovician in the vicinity of Cincinnati, Ohio. Cincinnati Museum of Natural History, 47 pp.

Cocks, L.R.M. 1968. Some strophomenacean brachiopods from the British Lower Silurian. British

Museum of Natural History (Geology), Bulletin 15: 283–324.

Cocks, L.R.M. 1990. Case 2747, *Strophomena* de Blainville, 1825 (Brachiopoda): proposed designation of *Leptaena planumbona* Hall, 1847 as the type species. Bulletin of Zoological Nomenclature, **47**: 274–276.

Cocks, L.R.M., and Rong, J.-y. 1988. A review of the Late Ordovician *Foliomena* brachiopod fauna with new data from China, Wales and Poland. Palaeontology, **31**: 53–67.

Cocks, L.R.M., and Rong, J.-y. 1989. Classification and review of the brachiopod superfamily Plectambonitacea. British Museum of Natural History (Geology), Bulletin 45: 77–163.

Cocks, L.R.M., and Rong, J.-y. 2000. Order Strophomenida. *In* Treatise on Invertebrate Paleontology, Part H, Brachiopoda (revised), Volumes 2 & 3: Linguliformea, Craniiformea, and Rhynchonelliformea (part). *Edited by* R.L. Kaesler. Geological Society of America and University of Kansas Press, Lawrence, pp. 216–347.

Conrad, T.A. 1842. Observations on the Silurian and Devonian systems of the U.S., with descriptions of new organic remains. Journal of the Academy of Natural Sciences of Philadelphia, **8**: 228–280.

Cooper, G.A. 1944. Phylum Brachiopoda. *In* Index fossils of North America. *Edited by* H.W. Shimer and R.R. Shrock. MIT Press, Cambridge, pp. 277–365.

Cooper, G.A. 1956. Chazyan and related brachiopods. Smithsonian Miscellaneous Collections, Vol. 127, 1245 pp.

Cowan, J. 1971. Ordovician and Silurian stratigraphy of the Interlake area, Manitoba. *In* Geoscience studies in Manitoba. *Edited by* A.C. Turnock. Geological Association of Canada, Special Paper 9: 235–241.

Dalman, J.W. 1828. Uppställning och Beskrifning af de i sverige funne Terebratuliter. Kungliga Vetenskapsakademien Handlingar, **3**: 85–155.

Davidson, T. 1853. British fossil Brachiopoda. Vol. I. Introduction. Monographs of the Palaeontographical Society, 136 pp.

Davis, R.A. 1985. Cincinnati fossils, an elementary guide to the Ordovician rocks and fossils of the Cincinnati, Ohio, region. Cincinnati Museum of Natural History Popular Publication Series, Vol. 10, 60 pp.

de Blainville, H.M.D. 1824. Dictionnaire des sciences naturelles. 2nd edition. Paris, 324 pp.

de Verneuil, E. 1848. Note sur quelques brachiopodes de L'ile de Gothland. Bulletin de Société géologique de France, **2**: 339–347.

Dewing, K. 1995. Late Ordovician and Early Silurian strophomenid brachiopods from Anticosti Island, Québec. Ph.D. thesis, University of Western Ontario, London.

Dewing, K. 1999. Late Ordovician and Early Silurian strophomenid brachiopods of Anticosti Island, Québec, Canada. Paleontographica Canadiana, Vol. 17, 143 pp.

Dowling, D.B. 1895. Notes on the stratigraphy of the Cambro–Silurian rocks of eastern Manitoba. Ottawa Naturalist, **9**: 65–73.

Dowling, D.B. 1900 [volume cover dated 1898]. Report on the geology of the west shore and islands of Lake Winnipeg. Geological Survey of Canada, Annual Report (New Series) 11, 100 pp.

Elias, R.J. 1980. Borings in solitary rugose corals of the Selkirk Member, Red River Formation (late Middle or Upper Ordovician), southern Manitoba. Canadian Journal of Earth Sciences, **17**: 272–277.

Elias, R.J. 1981. Solitary rugose corals of the Selkirk Member, Red River Formation (late Middle or Upper Ordovician), southern Manitoba. Geological Survey of Canada, Bulletin 344, 53 pp.

Elias, R.J. 1982. Paleoecology and biostratinomy of solitary rugose corals in the Stony Mountain Formation (Upper Ordovician), Stony Mountain, Manitoba. Canadian Journal of Earth Sciences, **19**: 1582–1598.

Elias, R.J. 1983. Late Ordovician solitary rugose corals of the Stony Mountain Formation, southern Manitoba, and its equivalents. Journal of Paleontology, **57**: 923–956.

Elias, R.J. 1985. Solitary rugose corals of the Upper Ordovician Montoya Group, southern New Mexico and westernmost Texas. Paleontological Society, Memoir 16, 58 pp.

Elias, R.J. 1991. Environmental cycles and bioevents in the Upper Ordovician Red River – Stony Mountain solitary rugose coral province of North America. Geological Survey of Canada, Paper 90-9: 205–211.

Elias, R.J., Nowlan, G.S., and Bolton, T.E. 1988. Paleontology of the type section, Fort Garry Member, Red River Formation (Upper Ordovician), southern Manitoba. New Mexico Bureau of Mines and Mineral Resources, Memoir 44: 341–359.

Emmons, E. 1842. Geology of New York, part 2, comprising the survey of the second geological district. W.A. White and J. Visscher, Albany.

Ethington, R.L., and Furnish, W.M. 1960. Upper Ordovician conodonts from southern Manitoba. Journal of Paleontology, **34**: 265–274.

Fenton, C.L., and Fenton, M.A. 1923. Some Black River brachiopods from the Mississippi Valley. Proceedings of Iowa Academy of Sciences, **29**: 67–77 [for 1922].

Foerste, A.F. 1912. *Strophomena* and other fossils from Cincinnatian and Mohawkian horizons, chiefly in Ohio, Indiana, and Kentucky. Scientific Laboratory of Denison University, Bulletin 17: 17–173.

Foerste, A.F. 1917. The Richmondian faunas of Little Bay de Noquette, in northern Michigan. The Ottawa Naturalist, **31**: 93–103, 121–127.

Foerste, A.F. 1924. Upper Ordovician faunas of Ontario and Quebec. Geological Survey of Canada, Memoir 138, 255 pp.

Foerste, A.F. 1929. The cephalopods of the Red River Formation of southern Manitoba. Journal of the Scientific Laboratories, Denison University, **24**: 129–235.

Gray, J.E. 1848. On the arrangement of the Brachiopoda. Annals and Magazine of Natural History, Series 2, **2**: 435–440.

Haidl, F.M., Longman, M.W., Pratt, B.R., and Bernstein, L.M. 1997. Variations in lithofacies in Upper Ordovician Herald and Yeoman formations (Red River), North Dakota and southeastern Saskatchewan. CSPG–SEPM Core Conference, Williston Basin Core Symposium, pp. 5–39.

Hall, J. 1847. Descriptions of the organic remains of the lower division of the New York System. New York State Geological Survey, Palaeontology of New York, Vol. 1, 338 pp.

Hall, J. 1859. Catalogue of the species of fossils of New York. New York State Cabinet of Natural History, Annual Report 12, pp. 63–96.

Hall, J. 1860. Contributions to palaeontology, 1858 and 1859. 13th Annual Report of the Regents of the University of the State of New York, on the condition of the State Cabinet of Natural History, pp. 55–128.

Hall, J., and Clarke, J.M. 1892. An introduction to the study of the genera of Palaeozoic Brachiopoda. New York State Geological Survey, Palaeontology of New York, **8**(1): 1–367, pls. 1–20.

Hall, J., and Clarke, J.M. 1894. An introduction to the study of the genera of Palaeozoic Brachiopoda. New York State Geological Survey, Palaeontology of New York, **8**(2): 319–394, pls. 21–84.

Harper, C.W., Jr., Boucot, A.J., and Walmsley, V.G. 1969. The rhipidomellid brachiopod subfamilies Heterorthinae and Platyorthinae (new). Journal of Paleontology, **43**: 74–92.

Hein, F.J., and Nowlan, G.S. 1998. Regional sedimentology, conodont biostratigraphy and correlation of Middle Cambrian – Lower Ordovician (?) strata of the “Finnegan” and Deadwood formations, Alberta subsurface, Western Canada Sedimentary Basin. Canadian Petroleum Geology, Bulletin 46: 166–188.

Hendricks, M.L., Eisel, J.D., and Fischer, W. 1998. Deadwood and Winnipeg sandstone reservoirs, Newporte Field, Renville County, North Dakota. *In* 8th International Williston Basin Symposium, Core Workshop Volume. Saskatchewan, North Dakota & Montana Geological Societies, pp. 1–8.

Howe, H.J. 1965. Plectambonitacea, Strophomenacea, and Atrypacea from the Montoya Group (Ordovician) of Trans-Pecos Texas. Journal of Paleontology, **39**: 647–656.

Howe, H.J. 1966. Orthacea from the Montoya Group (Ordovician) of Trans-Pecos Texas. Journal of Paleontology, **40**: 241–257.

Howe, H.J. 1967. Rhynchonellacea from the Montoya Group (Ordovician) of Trans-Pecos Texas. Journal of Paleontology, **41**: 845–860.

Howe, H.J. 1969. Rhynchonellacean brachiopods from the Richmondian of Tennessee. Journal of Paleontology, **43**: 1331–1350.

Howe, H.J. 1972. Morphology of the brachiopod genus *Thaerodonta*. Journal of Paleontology, **46**: 440–446.

Howe, H.J. 1979. Middle and Late Ordovician plectambonitacean, rhynchonellacean, syntrophiacean, and trimerellacean brachiopods from Kentucky. United States Geological Survey, Professional Paper 1066C: 1–18.

Howe, H.J. 1988. Articulate brachiopods from the Richmondian of Tennessee. Journal of Paleontology, **62**: 204–218.

Howe, H.J., and Reso, A. 1967. Upper Ordovician brachiopods from the Ely Springs Dolomite in southwestern Nevada. Journal of Paleontology, **41**: 351–363.

Hussey, R.C. 1926. The Richmondian Formation of Michigan. Contributions from the Museum of Geology, University of Michigan, **2**(8): 113–188.

James, U.P. 1874. Descriptions of new species of Brachiopoda, from the Lower Silurian rocks – Cincinnati Group. Cincinnati Quarterly Journal of Science, **1**: 19–22.

James, U.P. 1881. Contributions to palaeontology: fossils of the Lower Silurian Formation of Ohio, Indiana, and Kentucky. Palaeontologist, **5**: 33–44.

Jin, J. 1989. Late Ordovician – Early Silurian rhynchonellid brachiopods from Anticosti Island, Quebec. Biostratigraphie du Paléozoique, Vol. 10, 127 pp.

Jin, J. 1996. Ordovician (Llanvirn–Ashgill) rhynchonellid brachiopod biogeography. *In* Brachiopods. *Edited by* P. Copper and J. Jin. J. Balkema, Rotterdam, pp. 123–132.

Jin, J. 1999. Evolution and extinction of the Late Ordovician epicontinental brachiopod fauna of North America. Acta Universitatis Carolinae-Geologica, **43**: 203–206.

Jin, J. 2001. Evolution and extinction of the North American *Hiscobeccus* brachiopod Fauna during the Late Ordovician. Canadian Journal of Earth Sciences, **38**: 143–151.

Jin, J., and Chatterton, B.D.E. 1997. Late Ordovician – Silurian articulate brachiopods and biostratigraphy of the Avalanche Lake area, southwestern District of Mackenzie, Canada. Palaeontographica Canadiana, Vol. 13, 167 pp.

Jin, J., and Copper, P. 1997. *Parastrophinella* (Brachiopoda): its paleogeographic significance at the Ordovician/Silurian boundary. Journal of Paleontology, **71**: 369–380.

Jin, J., and Copper, P. 1998. *Kulumbella* and *Microcardinalia* (*Chiastodoca*) new subgenus, Early Silurian divaricate stricklandiid brachiopods from Anticosti Island, Eastern Canada. Journal of Paleontology, **72**: 441–453.

Jin, J., and Copper, P. 2000. Late Ordovician and Early Silurian pentamerid brachiopods from Anticosti Island, Québec. Palaeontographica Canadiana, Vol. 18, 139 pp.

Jin, J., and Lenz, A.C. 1992. An Upper Ordovician *Lepidocyclus–Hypsiptycha* fauna (rhynchonellid Brachiopoda) from the Mackenzie Mountains, Northwest Territories, Canada. Palaeontographica (A), **224**: 133–158.

Jin, J., and Norford, B.S. 1996. Upper Middle Ordovician (Caradoc) brachiopods from the Advance Formation, northern Rocky Mountains, British Columbia. Geological Survey of Canada, Bulletin 491: 20–77.

Jin, J., and Zhan, R. 2000. Evolution of the Late Ordovician orthid brachiopod *Gnamptorhynchos* Jin, 1989 from *Platystrophia* King, 1850 in North America. Journal of Paleontology, **74**: 983–991.

Jin, J., Caldwell, W.G.E., and Norford, B.S. 1989. Rhynchonellid brachiopods from the Upper Ordovician – Lower Silurian Beaverfoot and Nonda formations of the Rocky Mountains, British Columbia. Contributions to Canadian Paleontology, Geological Survey of Canada, Bulletin 396: 21–59.

Jin, J., Caldwell, W.G.E., and Norford, B.S. 1993. Early Silurian brachiopods and biostratigraphy of the Hudson Bay Lowlands, Manitoba, Ontario, and Quebec. Geological Survey of Canada, Bulletin 457, 221 pp.

Jin, J., Caldwell, W.G.E., and Norford, B.S. 1995. Late Ordovician brachiopods *Rafinesquina lata* Whiteaves, 1896 and *Kjaerina hartae* n. sp. from southern Manitoba and the Hudson Bay Lowlands. Canadian Journal of Earth Sciences, **32**: 1255–1266.

Jin, J., Caldwell, W.G.E., and Norford, B.S. 1997. Late Ordovician brachiopods and biostratigraphy of the Hudson Bay Lowlands, northern Manitoba and Ontario. Geological Survey of Canada, Bulletin 513, 115 pp.

Jin, J., Haidl, F.M., Bezys, R.K., and Gerla, G. 1999. The Early Silurian *Virgiana* brachiopod beds in the northeastern Williston Basin, Manitoba and Saskatchewan. Summary of Investigations 1999, Saskatchewan Geological Survey, **1**: 1–11.

Johnson, M.E., and Lescinsky, H.L. 1986. Depositional dynamics of cyclic carbonates from Interlake Group (Lower Silurian) of the Williston Basin. Palaios, **1**: 111–121.

Jones, O.T. 1928. *Plectambonites* and some allied genera. Geological Survey of Great Britain, Palaeontology, Memoir 1: 367–527.

Kaplan, P., Leighton, L.R., and Hermoyian, C.S. 1999. Rong is right: new strophomenoid classification accommodates cladistic and stratocladistic methods. Geological Society of America, 33rd Annual Meeting North-Central Section, Abstracts with Programs. p. A-26.

Kendall, A.C. 1976. The Ordovician carbonate succession (Bighorn Group) of southeastern Saskatchewan. Saskatchewan Geological Survey, Report 180, 185 pp.

Kendall, A.C. 1977. Origin of dolomite mottling in Ordovician limestones from Saskatchewan and Manitoba. Bulletin of Canadian Petroleum Geology, **25**: 480–504.

Kessler, L.G. 1991. Subsurface controlled stratigraphic sequences and the origin of shelf sand ridges, Winnipeg Group (Middle Ordovician) Manitoba, Saskatchewan, North Dakota. Sixth International Williston Symposium, Saskatchewan Geological Society Special Publication, **11**: 1–13.

King, W. 1846. Remarks on certain genera belonging to the class Palliobranchiata. Annals and Magazine of Natural History, **18**: 26–42.

Klenina, L.N., Nikitin, I.F., and Popov, L.E. 1984. Brakhiopody i biostratigrafiya srednego i verkhnego ordovika khrebta Chingiz. Alma-Ata, Nauka, pp. 6–125.

Kuhn, O. 1949. Lehrbuch der Paläozoologie. Schweizerbartsche Verlagsbuchhandlung, Stuttgart.

Kulkov, N.P., and Severgina, L.G. 1989. Stratigrafiya i brakhiopody ordovika i nizhnego silura Gornogo Altaya. Trudy Instituta Geologii i Geofiziki, Akademiya Nauk SSSR, Sibirskoye Otdeleniye, Vol. 717, 223 pp.

Le Fèvre, J., Barnes, C.R., and Tixier, M. 1976. Paleoecology of Late Ordovician and Early Silurian conodontophorids, Hudson Bay Basin. Geological Association of Canada, Special Paper 15: 69–89.

Leith, E.I. 1952. Schizocoralla from the Ordovician of Manitoba. Journal of Paleontology, **26**: 789–796.

Lockley, M.G. 1983. A review of brachiopod dominated palaeocommunities from the type Ordovician. Palaeontology, **26**: 111–145.

Logan, W.E. 1863. Geology of Canada. Geological Survey of Canada, report of progress from its commencement to 1863. Dawson Brothers, Montreal.

Longman, M.W., and Haidl, F.M. 1996. Cyclic deposition and development of porous dolomites in the Upper Ordovician Red River Formation, Williston Basin. *In* Paleozoic systems of the Rocky Mountain Region. *Edited by* M.W. Longman and M.D. Sonnenfeld. Rocky Mountain Section, Society of Economic Paleontologists and Mineralogists, pp. 29–46.

Longman, M.W., Bogle, R., and Single, E.L. 1998. Lantry Field, South Dakota: an odd Red River Reservoir on the southeast flank of the Williston Basin. Eighth International Williston Basin Symposium. Saskatchewan Geological Society Special Publication, **13**: 14–23.

Macomber, R.W. 1970. Articulate brachiopods from the upper Bighorn Formation (Late Ordovician) of Wyoming. Journal of Paleontology, **44**: 416–450.

McCabe, H.R. 1971. Stratigraphy of Manitoba, an introduction and review. Geological Association of Canada, Special Paper 9: 167–187.

McCabe, H.R., and Bannatyne, B.B. 1970. Paleozoic and Mesozoic of the Dawson Bay area and the Manitoba Escarpment, Manitoba. Geological Association of Canada and Mineralogical Association of Canada. Guidebook for Field Trip No. 6.

McCabe, H.R., and Barchyn, D. 1982. Paleozoic stratigraphy of southwestern Manitoba. Geological Association of Canada and Mineralogical Association of Canada, Guidebook for Field Trip No. 10.

Meek, F.B. 1873. Descriptions of invertebrate fossils of the Silurian and Devonian systems, Volume 1, Part 2, Palaeontology. Ohio Geological Survey.

Miller, S.A. 1875. Monograph of the class Brachiopoda of the Cincinnati Group. Cincinnati Quarterly Journal of Science, **2**: 6–62.

Mitchell, W.I. 1977. The Ordovician Brachiopoda from Pomeroy, Co. Tyrone. Palaeontological Society Monograph, Vol. 130, 138 pp.

Muir-Wood, H., and Williams, A. 1965. Strophomenida. *In* Treatise on invertebrate paleontology, Part H, Volume 1. *Edited by* R.C. Moore. Geological Society of America and University of Kansas Press, Lawrence, pp. 361–521.

Nelson, S.J. 1959. Arctic Ordovician fauna: an equatorial assemblage? Journal of the Alberta Society of Petroleum Geologists, **7**: 45–47, 53.

Nelson, S.J. 1963. Ordovician paleontology of the northern Hudson Bay Lowland. Geological Society of America, Memoir 90, 152 pp.

Nelson, S.J. 1964. Ordovician stratigraphy of northern Hudson Bay Lowland, Manitoba. Geological Survey of Canada, Bulletin 108, 36 pp.

Nettelroth, H. 1889. Kentucky fossil shells, a monograph of the fossil shells of the Silurian and Devonian rocks of Kentucky. Kentucky Geological Survey.

Neuman, R.B., and Bruton, D.L. 1974. Early Middle Ordovician fossils from the Hølonda area, Trondheim region, Norway. Norsk Geologisk Tidsskrift, **54**: 69–115.

Nikitin, I.F., and Popov, L.E. 1985. Ordoviskie Strofomenidy (Brakhiopody) Severnogo Priishimia (Tsentralnyi Kazakhstan). Ezhegodnik Vsesoyuznogo Paleontologischeskogo Obshchestva, **28**: 34–49.

Nikitin, I.F., Popov, L.E., and Rukavishnikova, T.B. 1980. Brachiopoda. *In* Granitsa Ordovika i Silura v Kazakhstane. *Edited by* M.K. Apollonov, S.M. Bandaletov, and I.F. Nikitin. Nauka KazSSR, Alma-Ata, pp. 37–73.

Nikitin, I.F., Popov, L.E., and Holmer, L.E. 1996. Late Ordovician brachiopod assemblage of Hiberno–Salarian type from Central Kazakhstan. GFF, **118**: 84–96.

Norford, B.S., Haidl, F.M., Bezys, R.K., Cecile, M.P., McCabe, H.R., and Paterson, D.F. 1994. Middle Ordovician to Lower Devonian strata of the Western Canada Sedimentary Basin. *In* Geological Atlas of the Western Canada Sedimentary Basin. *Compiled by* G. Mossop and I. Shetsen. Canadian Society of Petroleum Geologists and Alberta Research Council, pp. 109–127.

Norford, B.S., Nowlan, G.S., Haidl, F.M., and Bezys, R.K. 1998. The Ordovician–Silurian boundary interval in Saskatchewan and Manitoba. Eighth International Williston Basin Symposium, Saskatchewan Geological Society, Special Publication, **13**: 27–45.

Nowlan, G.S., and Haidl, F.M. 1999. New conodont data from the Ordovician–Silurian boundary interval in southeastern Saskatchewan. Saskatchewan Geological Survey, Summary of Investigations 1999, **1**: 12–16.

Oberg, R. 1966. Winnipeg conodonts from Manitoba. Journal of Paleontology, **40**: 130–147.

Okulitch, V.J. 1943. The Stony Mountain Formation of Manitoba. Transactions of the Royal Society of Canada, 3rd Series, **37**: 59–74.

Öpik, A.A. 1930. Brachiopoda Protremata der estländischen Ordovizischen Kukruse-Stufe. Tartu University, Acta & Commentationes, Ser. A, Vol. 17, 262 pp.

Öpik, A.A. 1934. Über Klitamboniten. Tartu University, Acta & Commentationes, Ser. A, Vol. 26(3), 239 pp.

Parks, W.A. 1917. Palaeozoic fossils from a region southwest of Hudson Bay. A description of fossils collected by Joseph B. Tyrrell, Esq., F.R.S.C., in the District of Patricia, Ontario, and northern Manitoba during the summer of 1912. Transactions of the Royal Canadian Institute, **11**: 3–95.

Percival, I.G. 1979. Late Ordovician articulate brachiopods from Gunningbland, central western New South Wales. Proceedings of the Linnean Society of New South Wales, **103**: 175–187.

Perkins, R.D., and Halsey, S.D. 1971. Geologic significance of microboring fungi and algae in Carolina shelf sediments. Journal of Sedimentary Petrology, **41**: 843–853.

Perkins, R.D., and Tsentas, C.I. 1976. Microbial infestation of carbonate substrates planted on the St. Croix shelf, West Indies. Geological Society of America, Bulletin 87: 1615–1628.

Petersen, C.G.J. 1914. Valuation of the Sea II, The animal communities of the sea bottom and their importance for marine zoogeography. Report on the Danish Biological Station to the Board of Agriculture (Ministry of Fisheries) Copenhagen, **21**: 3–44.

Pickerill, R.K., and Brenchley, P.J. 1979. Caradoc marine benthic communities of the south Berwyn Hills, North Wales. Palaeontology, **22**: 229–264.

Pope, J.K. 1976. Comparative morphology and shell histology of the Ordovician Strophomenacea (Brachiopoda). Palaeontographica Americana, **8**(49): 129–213.

Pope, J.K., and Martin, W.D. 1977. Field guide book to the biostratigraphy and paleoenvironments of the Cincinnatian Series of southwestern Indiana. Society of Economic Paleontologists and Mineralogists. Great Lakes Section, 7th Annual Conference, October 1977, Oxford, Ohio.

Potter, A.W., and Boucot, A.J. 1992. Middle and Late Ordovician brachiopod benthic assemblages of North America. *In* Global perspectives on Ordovician geology. *Edited by* B.D. Webby and J.R. Laurie. Balkema, Rotterdam, pp. 307–323.

Pratt, B.R., Bernstein, L.M., Kendall, A.C., and Haidl, F.M. 1996. Occurrence of reefal facies in Red River strata (Upper Ordovician), subsurface Saskatchewan. Summary of Investigations 1996, Saskatchewan Geological Survey, Saskatchewan Energy and Mines, Miscellaneous Report 96-4: 147–152.

Reed, F.R.C. 1917. The Ordovician and Silurian Brachiopoda of the Girvan District. Transactions of the Royal Society of Edinburgh, **51**: 795–998.

Rong, J.-y. 1984. Ecostratigraphic evidence of the Upper Ordovician regressive sequence and the effect of the glaciation. Journal of Stratigraphy, **8**: 19–29 [in Chinese].

Rong, J.-y. 1986. Ecostratigraphy and community analysis of the Late Ordovician and Silurian in Southern China. *In* Selected paper collections of the annual symposium of the thirteenth and fourteenth committee of the Palaeontological Society of China. Anhui Science and Technology Press, Hefei, pp. 1–24 [in Chinese with English summary].

Rong, J.-y., and Cocks, L.R.M. 1994. True *Strophomena* and a revision of the classification and evolution of strophomenoid and 'strophodontoid' brachiopods. Palaeontology, **37**: 651–694.

Rong, J.-y., and Harper, D.A.T. 1999. Brachiopod survival and recovery from the latest Ordovician mass extinctions in South China. Geological Journal, **34**: 321–348.

Rong, J.-y., and Zhan, R.-b. 1995. On the *Foliomena* fauna (Late Ordovician brachiopods). Chinese Science Bulletin, **40**: 928–931 [in Chinese].

Rong, J.-y., and Zhan, R.-b. 1996. Distribution and ecological evolution of the *Foliomena* Fauna (Late Ordovician brachiopods). *In* Centennial memorial volume of Prof. Sun Yunzhu: palaeontology and stratigraphy. *Edited by* Wang Hongzhen and Wang Xunlian. China University of Geosciences Press, Beijing, pp. 90–97.

Rong, J.-y., Li, R.-y., and Kulkov, N.P. 1995. Biogeographic analysis of Llandovery brachiopods from Asia with a recommendation of use of affinity indices. Acta Palaeontologica Sinica, **34**: 428–453.

Rong, J.-y., Zhan, R.-b., and Harper, D.A.T. 1999. Late Ordovician (Caradoc–Ashgill) brachiopod faunas with *Foliomena* based on data from China. Palaios, **14**: 412–431.

Ross, R.J., Jr., 1957. Ordovician fossils from wells in the Williston Basin, eastern Montana. United States Geological Survey, Bulletin 1021-M: 439–506.

Roy, S.K. 1941. The Upper Ordovician fauna of Frobisher Bay, Baffin Land. Field Museum of Natural History (Geology), Memoir 2, 212 pp.

Salmon, E.S. 1942. Mohawkian Rafinesquinae. Journal of Paleontology, **16**: 564–603.

Sardeson, F.W. 1892. The range and distribution of the Lower Silurian fauna of Minnesota with descriptions of some new species. Minnesota Academy of Natural Sciences, Bulletin 3: 326–343.

Schmidt, H., and McLaren, D.J. 1965. Paleozoic Rhynchonellacea. *In* Treatise on invertebrate paleontology, Part H, Volume 2. *Edited by* R.C. Moore. Geological Society of America and University of Kansas Press, Lawrence, pp. 552–597.

Schuchert, C. 1893. A classification of the Brachiopoda. American Geologist, **11**: 141–167.

Schuchert, C. 1913. Class 2. Brachiopoda. *In* Textbook of palaeontology, Volume 1, 2nd edition. *By* K.A. von Zittel. *Translated and edited by* C.R. Eastman. Macmillan, London, pp. 355–420.

Schuchert, C., and Cooper, G.A. 1931. Synopsis of the brachiopod genera of the suborders Orthoidea and Pentameroidea, with notes on the Telotremata. American Journal of Science, **20**: 241–251.

Schuchert, C., and Cooper, G.A. 1932. Brachiopod genera of the Suborders Orthoidea and Pentameroidea. Peabody Museum of Natural History, Memoir 4, Part 1, 270 pp.

Schuchert, C., and LeVene, C.M. 1929. Brachiopoda (Generum et genotyporum index et bibliographia). Fossilium Catalogus, 1 Animalia (42). Berlin, pp. 1–140.

Scotese, C.R., and McKerrow, W.S. 1990. Revised world maps and introduction. *In* Paleozoic paleogeography and biogeography. *Edited by* W.S. McKerrow and C.R. Scotese. Geological Society (London), Memoir 12: 1–21.

Shaler, N.S. 1865. List of the Brachiopoda from the island of Anticosti sent by the Museum of Comparative Zoology to different institutions for exchange for other specimens, with annotations. Harvard University, Museum of Comparative Zoology, Bulletin 1, pp. 61–70.

Sheehan, P.M., and Coorough, P.J. 1990. Brachiopod zoogeography across the Ordovician–Silurian extinction event. Geological Society (London), Memoir 12: 181–187.

Sinclair, G.W. 1959. Succession of Ordovician rocks in southern Manitoba. Geological Survey of Canada, Paper 59-5, 9 pp.

Sinclair, G.W., and Leith, E.I. 1958. New name for an Ordovician shale in Manitoba. Journal of Paleontology, **32**: 243–244.

Sloss, L.L. 1963. Sequences in the Cratonic interior of North America. Geological Society of America Bulletin 74: 93–111.

Sloss, L.L. 1988. Tectonic evolution of the craton in Phanerozoic time. *In* Sedimentary Cover – North American Craton: U.S. *Edited by* L.L. Sloss. The Geology of North America, Vol. D-2, pp. 25–51.

Smith, D.L. 1963. A lithologic study of the Stony Mountain and Stonewall Formations in southern Manitoba. M.Sc. thesis, University of Manitoba, Winnipeg.

Sokolskaya, A.N. 1960. Otryad Strophomenida. *In* Mshanki, Brakhiopody, Osnovy Paleontologii. *Edited by* T.G. Sarycheva. Izdatelstvo Akademii Nauk SSSR, Moscow, pp. 206–220.

Spjeldnaes, N. 1957. The Middle Ordovician of the Oslo region, Norway, 8. Brachiopoda of the Suborder Strophomenoidea. Norsk Geologisk Tidsskrift, Vol. 37, 214 pp.

Stearn, C.W. 1956. Stratigraphy and palaeontology of the Interlake Group and Stonewall Formation of southern Manitoba. Geological Survey of Canada, Memoir 281, 162 pp.

Sweet, W.C. 1979. Late Ordovician conodonts and biostratigraphy of the western Midcontinent Province. Brigham Young University, Geology Studies, **26**(3): 45–85.

Sweet, W.C. 1982. Conodonts from the Winnipeg Formation (Middle Ordovician) of the Northern Black Hills, South Dakota. Journal of Paleontology, **56**: 1029–1049.

Thorson, G. 1957. Bottom communities. *In* Treatise on marine ecology and paleoecology, Volume 1, Ecology. *Edited by* J.W. Hedgpeth. Geological Society of America, Memoir 67: 461–534.

Troedsson, G.T. 1928. On the Middle and Upper Ordovician faunas of northern Greenland, Part 2. Meddelelser om Grønland, Vol. 72, 197 pp.

Twenhofel, W.H. 1928. Geology of Anticosti Island. Geological Survey of Canada, Memoir 154, 481 pp.

Twenhofel, W.H., et al. 1954. Correlation of the Ordovician formations of North America. Geological Society of America, Bulletin 65: 247–298.

Ulrich, E.O. 1889. On some Polyzoa (Bryozoa) and Ostracoda from the Cambro–Silurian rocks of Manitoba. Geological and Natural History Survey of Canada, Contributions to the Micro-paleontology of the Cambro–Silurian Rocks of Canada, **2**(4): 28–57.

Ulrich, E.O., and Cooper, G.A. 1938. Ozarkian and Canadian Brachiopoda. Geological Society of America, Special Paper 13, 323 pp.

von Schlotheim, E.F. 1820. Die Petrefactenkunde auf ihrem jetzigen Standpunkte durch die Beschreibung einer Sammlung versteinerter unf fossiler Uberreste der Thier-und-Pfanzenreiches der Vorwelt erlaütert. I-LXII, 437 pp.

Wang, Y. 1949. Maquoketa Brachiopoda of Iowa. Geological Society of America, Memoir 42, 55 pp.

Weiss, M.P. 1955. Some Ordovician brachiopods from Minnesota and their stratigraphic relations. Journal of Paleontology, **29**: 759–774.

Westrop, S.R., and Ludvigsen, R. 1983. Systematics and paleoecology of Upper Ordovician trilobites from the Selkirk Member of the Red River Formation, southern Manitoba. Manitoba Department of Energy and Mines, Mineral Resources Division, Geological Report GR82-2, 51 pp.

White, C.A. 1880. Palaeontology, fossils of the Indiana rocks. Indiana Bureau of Statistics and Geology, 2nd Annual Report, pp. 471–544.

Whiteaves, J.F. 1880. On some Silurian and Devonian fossils from Manitoba and the valleys of the Nelson and Churchill rivers, for the most part collected by Dr. R. Bell in the summer of 1879. Geological Survey of Canada, Report of Progress for 1878–79, Part C, Appendix 1, pp. 45–51.

Whiteaves, J.F. 1895. Systematic list, with references, of the fossils of the Hudson River or Cincinnati formation at Stony Mountain, Manitoba. Geological Survey of Canada, Palaeozoic Fossils, **3**(2): 111–128.

Whiteaves, J.F. 1896. Descriptions of eight new species of fossils from the (Galena) Trenton limestones of Lake Winnipeg and the Red River Valley. Canadian Record of Science, **6**: 387–397.

Whiteaves, J.F. 1897. The fossils of the Galena–Trenton and Black River formations of Lake Winnipeg and its vicinity. Geological Survey of Canada, Palaeozoic Fossils, **3**(3): 129–242.

Whitfield, R.P. 1878. Preliminary descriptions of new species of fossils from the lower geological formations of Wisconsin. Annual Report of the Wisconsin Geological Survey for the Year 1877, pp. 50–89.

Whitfield, R.P. 1882. Palaeontology. Wisconsin Geological Survey, Geology of Wisconsin, **4**: 161–363.

Williams, A., and Harper, D.A.T. 2000. Order Orthida. *In* Treatise on invertebrate paleontology, Part H,

Brachiopoda (revised), Volumes 2 & 3: Linguliformea, Craniiformea, and Rhynchonelliformea (part). *Edited by* R.L. Kaesler. Geological Society of America and University of Kansas Press, Lawrence, pp. 714–846.

Williams, A., and Wright, A.D. 1965. Orthida. *In* Treatise on invertebrate paleontology, Part H, Volume 1. *Edited by* R.C. Moore. Geological Society of America and University of Kansas Press, Lawrence, pp. 299–395.

Wilson, A.E. 1926. An Upper Ordovician fauna from the Rocky Mountains, British Columbia. Canada Department of Mines, Geological Survey, Bulletin 44 (Geological Series, No. 46), pp. 1–34.

Wilson, A.E. 1945. Strophomena and its homomorphs *Trigrammaria* and *Microtrypa* from the Ottawa Limestone of the Ottawa – St. Lawrence lowlands. Transactions of the Royal Society of Canada, Series 3, **39**(4): 121–150.

Winchell, N.H., and Schuchert, C. 1892. Preliminary descriptions of new Brachiopoda from the Trenton and Hudson River groups of Minnesota. American Geologist, **9**: 284–294.

Winchell, N.H., and Schuchert, C. 1893. The Lower Silurian Brachiopoda of Minnesota. The Geology of Minnesota, Minnesota Geological and Natural history Survey, Final Report, **3**(1): 333–374 [entire volume dated 1895].

Woodward, S.P. 1852. A manual of the Mollusca; or rudimentary treatise of recent and fossil shells. London, 486 pp.

Zhan, R.-b., and Cocks, L.R.M. 1998. Late Ordovician brachiopods from the South China Plate and their palaeogeographical significance. Special Papers in Palaeontology, Vol. 59, 70 pp.

Zhan, R.-b., and Li, R.-y. 1998. The discovery of *Altaethyrella* Severgina 1978 (Late Ordovician rhynchonelloid brachiopods) in China. Acta Palaeontologica Sinica, **37**: 194–211.

Zhan, R.-b., and Rong, J.-y. 1995. Distribution pattern of late Ordovician brachiopod communities in the Zhejiang–Jiangxi border region, E. China. Chinese Science Bulletin, **40**: 932–935 [in Chinese].

Ziegler, A.M. 1965. Silurian marine communities and their environmental significance. Nature, **207**: 270–272.

Ziegler, A.M., Cocks, L.R.M., and Bambach, R.K. 1968. The composition and structure of Lower Silurian marine communities. Lethaia, **1**: 1–27.

Appendix A

Ashgill brachiopod faunal data for paleobiogeographic analysis

1. Southern Manitoba (this study)

Order	Superfamily	Genera
Orthida (20.0%)	Orthoidea	*Dinorthis, Platystrophia*
	Dalmanelloidea	*Diceromyonia*
Strophomenida (46.6%)	Plectambonitoidea	*Thaerodonta*
	Strophomenoidea	*Strophomena, Holtedahlina, Tetraphalerella, Oepikina, Kjaerina, Megamyonia*
Pentamerida (6.7%)	Porambonitoidea	*Parastrophinella*
Rhynchonellida (26.7%)	Rhynchonelloidea	*Rhynchotrema, Lepidocyclus, Hypsiptycha, Hiscobeccus*

2. Hudson Bay Lowlands (Jin et al. 1997)

Order	Superfamily	Genera
Orthida (20.0%)	Orthoidea	*Dinorthis, Plaesiomys*
	Dalmanelloidea	*Diceromyonia*
Strophomenida (46.6%)	Plectambonitoidea	*Thaerodonta*
	Strophomenoidea	*Strophomena, Tetraphalerella, Oepikina, Kjaerina, Megamyonia, Leptaena*
Rhynchonellida (20.0%)	Rhynchonelloidea	*Hiscobeccus, Hypsiptycha, Rostricellula*
Atrypida (13.4%)	Zygospirioidea	*Zygospira, Catazyga*

3. Wyoming (Macomber 1970)

Order	Superfamily	Genera
Orthida (35.7%)	Orthoidea	*Hesperorthis, Dinorthis, Plaesiomys, Platystrophia*
	Dalmanelloidea	*Diceromyonia*
Strophomenida (28.6%)	Plectambonitoidea	*Thaerodonta*
	Strophomenoidea	*Strophomena, Oepikina, Megamyonia*
Rhynchonellida (21.4%)	Rhynchonelloidea	*Rhynchotrema, Hiscobeccus, Hypsiptycha*
Atrypida (14.3%)	Zygospirioidea	*Zygospira*
	Lissatrypoidea	*Cyclospira*

4. Iowa (Wang 1949)

Order	Superfamily	Genera
Orthida (38.9%)	Orthoidea	*Glyptorthis, Plaesiomys, Austinella, Platystrophia*
	Dalmanelloidea	*Diceromyonia, Dalmanella, Onniella*
Strophomenida (33.3%)	Plectambonitoidea	*Thaerodonta*
	Strophomenoidea	*Strophomena, Tetraphalerella, Oepikina, Holtedahlina, Megamyonia*
Pentamerida (5.6%)	Porambonitoidea	*Parastrophinella*
Rhynchonellida (16.7%)	Rhynchonelloidea	*Rhynchotrema, Hiscobeccus, Hypsiptycha, Lepidocyclus*
Atrypida (5.6%)	Zygospirioidea	*Zygospira*

5. Tennessee (Howe 1969, 1988)

Order	Superfamily	Genera
Orthida (58.8%)	Orthoidea	*Glyptorthis, Plaesiomys, Retrorsirostra, Austinella, Platystrophia, Hebertella, Pionodema*
	Dalmanelloidea	*Diceromyonia, Paucicrura, Onniella*
Strophomenida (23.5%)	Plectambonitoidea	*Thaerodonta*
	Strophomenoidea	*Strophomena, Tetraphalerella, Rafinesquina*
Rhynchonellida (11.8%)	Rhynchonelloidea	*Rhynchotrema, Lepidocyclus*
Atrypida (5.9%)	Zygospirioidea	*Zygospira*

6. Jiangshan–Changshan–Yushan area, South China (Zhan and Cocks 1998)

Order	Superfamily	Genera
Orthida (30.6%)	Orthoidea	*Plectorthis, Mimella, Skenidioides*
	Dalmanelloidea	*Epitomyonia, Wangyuella, Dedztina*
	Triplesioidea	*Triplesia, Oxoplecia*
Strophomenida (44.4%)	Plectambonitoidea	*Bimuria, Metambonites, Synambonites, Anoptambonites, Rongambonites, Kassinella, Sowerbyella, Rugosowerbyella*
	Strophomenoidea	*Strophomena, Holtedahlina, Fenomena, Tashanomena, Foliomena, Christiania*
Pentamerida (8.3%)	Camerelloidea	*Eostrophina*
	Pentameroidea	*Tcherskidium*
Rhynchonellida (2.8%)	Rhynchonelloidea	*Altaethyrella*
Atrypida (11.1%)	Zygospirioidea	*Antizygospira, Ovalospira*
	Atrypoidea	*Eospirigerina*
	Lissatrypoidea	*Cyclospira*
Spiriferida (2.8%)	Cyrtioidea	*Eospirifer*

7. Dulankara area, Kazakhstan (Nikitin et al. 1980; Klenina et al. 1984)

Order	Superfamily	Genera
Orthida (25.0%)	Orthoidea	*Dinorthis, Mimella, Schizophorella*
	Dalmanelloidea	*Dalmanella, Paucicrura*
	Triplesioidea	*Triplesia*
Strophomenida (37.5%)	Plectambonitoidea	*Leptellina, Sowerbyella, Rugosowerbyella, Viruella, Eoplectodonta, Kassinella*
	Strophomenoidea	*Strophomena, Oepikina, Christiania*
Pentamerida (8.3%)	Pentameroidea	*Tcherskidium, Prostricklandia*
Rhynchonellida (4.2%)	Rhynchonelloidea	*Altaethyrella*
Atrypida (16.7%)	Zygospirioidea	*Ovalospira, Zygospiraella*
	Atrypoidea	*Eospirigerina*
	Lissatrypoidea	*Cyclospira*
Spiriferida (4.2%)	Cyrtioidea	*Iliella*
Athyrida (4.2)	Athyridioidea	*Cryptothyrella*

8. Gorny Altai, Siberia (Kulkov and Severgina 1989)

Order	Superfamily	Genera
Orthida (42.3%)	Orthoidea	*Glyptorthis, Ptychopleurella, Austinella, Plaesiomys, Severginella, Schizophorella*
	Dalmanelloidea	*Salopina, Epitomyonia, Reuschella*
	Triplesioidea	*Triplesia, Oxoplecia*
Strophomenida (30.8%)	Plectambonitoidea	*Anoptambonites, Dulankarella, Diambonia, Sowerbyella, Bimuria*
	Strophomenoidea	*Strophomena, Glyptomena, Mjoesina*
Rhynchonellida (15.4%)	Rhynchonelloidea	*Altaethyrella, Rhynchotrema, Rostricellula, Salairella*
Atrypida (11.5%)	Zygospirioidea	*Catazyga, Zygospira*
	Atrypoidea	*Eospirigerina*

Appendix B

Locality data

1. GSC localities

GSC loc. O-27185 (Field No. 55-51), Sinclair's collection (1955), Gunn Member, Stony Mountain Formation. City of Winnipeg Quarry, Stony Mountain, 50°05'34"N, 97°12'25"W. *Dinorthis occidentalis*, *Diceromyonia storeya*, *Megamyonia nitens*, *Oepikina limbrata*, *Hypsiptycha occidens*.

GSC loc. O-27186 (Field No. 55-52), Sinclair's collection (1955), Penitentiary Member, Stony Mountain Formation. City of Winnipeg Quarry, Stony Mountain, 50°05'34"N, 97°12'25"W. *Dinorthis occidentalis*, *Diceromyonia storeya*, *Strophomena vetusta*, *Nasutimena fluctuosa*, *Megamyonia nitens*, *Oepikina limbrata*, *Hypsiptycha occidens*, *Hiscobeccus gigas*.

GSC loc. O-37130 (Field No. SB-57-19), Sinclair's collection (1957), Gunn Member, Stony Mountain Formation. Stony Mountain Penitentiary, 50°04'N, 97°12'W. *Dinorthis occidentalis*, *Diceromyonia storeya*, *Strophomena planumbona*, *Strophomena vetusta*, *Nasutimena fluctuosa*, *Megamyonia nitens*, *Oepikina limbrata*, *Hypsiptycha occidens*, *Hiscobeccus gigas*.

GSC loc. C-205926, Gunn Member, Stony Mountain Formation. Stony Mountain, 50°04'40"N, 97°12'40"W. *Dinorthis occidentalis*, *Diceromyonia storeya*, *Strophomena vetusta*, *Nasutimena fluctuosa*, *Oepikina limbrata*, *Hypsiptycha occidens*, *Hiscobeccus gigas*.

GSC loc. C-205928, Gunn Member, Stony Mountain Formation. Stony Mountain, 50°04'52"N, 97°12'40"W. *Dinorthis occidentalis*, *Diceromyonia storeya, Oepikina limbrata, Hypsiptycha occidens, Hiscobeccus gigas*.

GSC loc. C-205930, Gunn Member, Stony Mountain Formation. Stony Mountain, 50°04'56", 97°12'42". *Dinorthis occidentalis, Diceromyonia storeya, Nasutimena fluctuosa, Oepikina limbrata, Hypsiptycha occidens, Hiscobeccus gigas*.

GSC loc. C-205931, Gunn Member, Stony Mountain Formation. City of Winnipeg Quarry, 50°05'34"N, 97°12'25"W. *Dinorthis occidentalis, Diceromyonia storeya, Nasutimena fluctuosa, Megamyonia nitens, Oepikina limbrata, Hiscobeccus gigas*.

GSC loc. C-205933, Penitentiary Member, Stony Mountain Formation. City of Winnipeg Quarry, 50°05'34"N, 97°12'35"W. *Dinorthis occidentalis, Diceromyonia storeya*.

GSC loc. C-205935, Selkirk Member, Red River Formation, Garson Quarry, 50°05'N, 96°42'W. *Dinorthis occidentalis, Gnamptorhynchos manitobensis, Diceromyonia storeya, Thaerodonta clarksvillensis, Strophomena vetusta, Nasutimena fluctuosa, Nasutimena undulosa, Tetraphalerella neglecta, Tetraphalerella churchillensis, Kjaerina hartae, Megamyonia nitens, Oepikina lata, Parastrophinella cirrita, Rhynchotrema iowense, Lepidocyclus laddi, Hypsiptycha anticostiensis, Hypsiptycha occidens, Hiscobeccus capax*.

GSC loc. 55-60, Penitentiary Member, Stony Mountain Formation. Stony Mountain, 50°04'N, 97°12'W. *Dinorthis occidentalis, Diceromyonia storeya, Thaerodonta clarksvillensis, Strophomena vetusta, Nasutimena fluctuosa, Oepikina limbrata, Hypsiptycha occidens, Hiscobeccus gigas*.

2. Manitoba Museum of Man and Nature (MMMN) collections

MMMN I-94, Selkirk Member, Red River Formation. Garson, 50°05'00"N, 96°42'00"W. *Kjaerina hartae*.

MMMN I-185, Selkirk Member, Red River Formation. Garson, 50°05'00"N, 96°42'00"W. *Tetraphalerella churchillensis*.

MMMN I-377, Selkirk Member, Red River Formation. Gillis Quarry, Garson, 50°05'00"N, 96°42'00"W. *Oepikina lata*.

MMMN I-423, Selkirk Member, Red River Formation. Garson, 50°05'00"N, 96°42'00"W. *Strophomena vetusta.*

MMMN I-686, Selkirk Member, Red River Formation. Garson Quarry, 50°05'00"N, 96°42'00"W. *Kjaerina hartae.*

MMMN I-900, Selkirk Member, Red River Formation. Garson Quarry, 50°05'00"N, 96°42'00"W. *Diceromyonia storeya.*

MMMN I-1654, Selkirk Member, Red River Formation. Rubble heaps of Gillis Quarry, Garson, 50°05'00"N, 96°42'00"W. *Oepikina lata.*

MMMN I-1678, Selkirk Member, Red River Formation. Rubble heaps of Gillis Quarry, Garson, 50°04'26"N, 96°42'14"W. *Tetraphalerella churchillensis.*

MMMN I-1871, Selkirk Member, Red River Formation. Rubble heaps of Gillis Quarry, Garson, 50°05'00"N, 96°42'00"W. *Oepikina lata.*

MMMN I-2150, Selkirk Member, Red River Formation. Garson, 50°05'00"N, 96°42'00"W. *Oepikina lata.*

MMMN I-2152, Selkirk Member, Red River Formation. Garson, 50°05'00"N, 96°42'00"W. *Kjaerina hartae.*

MMMN I-2153, Selkirk Member, Red River Formation. Garson, 50°05'00"N, 96°42'00"W. *Nasutimena fluctuosa.*

MMMN I-2164, Selkirk Member, Red River Formation. South rubble heap, Gillis Quarry, Garson, 50°05'00"N, 96°42'00"W. *Gnamptorhynchos manitobensis.*

MMMN I-2165, Selkirk Member, Red River Formation. In the northwest corner of the Gillis Quarry, Garson, 50°05'00"N, 96°42'00"W. *Diceromyonia storeya.*

MMMN I-2415, Dog Head Member, Red River Formation. Quarry about 1 km southwest of Pine Dock, 51°37'30"N, 96°49'00"W. *Thaerodonta clarksvillensis.*

MMMN I-2416, upper Dog Head Member, Red River Formation. Quarry about 1 km southwest of Pine Dock, 51°37'30"N, 96°49'00"W. *Nasutimena fluctuosa.*

MMMN I-2418, upper Dog Head Member, Red River Formation. West of Hecla, north of main road, Old Quarry, 51°07'00"N, 96°56'00"W. *Diceromyonia storeya.*

MMMN I-2419, Dog Head Member, Red River Formation. North end of Hecla Island, West Quarry, 51°11'30"N, 96°41'00"W. *Thaerodonta clarksvillensis.*

MMMN I-2472, Selkirk Member, Red River Formation. Gillis Quarry, large heap of rubble, Garson, 50°04'26"N, 96°42'14"W. *Thaerodonta clarksvillensis.*

MMMN I-2543, Fort Garry Member, Red River Formation. Wright site, west bank of Red River about 300 m south of Parkes Creek (at the base of the bank), 50°01'30"N, 97°02'11"W. *Holtedahlina paraprostrata.*

MMMN I-2653, Hecla Beds?, Red River Formation. Victoria Beach, 50°42'08"N, 96°33'46"W. *Diceromyonia storeya.*

MMMN I-2680, Selkirk Member, Red River Formation. Old quarry section south of Koostatak, Koostatak Dump, 51°24'51"N, 97°21'19"W. *Hypsiptycha occidens.*

MMMN I-2766, lower Selkirk Member, Red River Formation. Henry Kazina Quarry, northwest of rubble heap, north of Tyndall, 50°06'54"N, 96°40'24"W. *Thaerodonta clarksvillensis.*

MMMN I-2767, lower Selkirk Member, Red River Formation. Northwest of rubble heap, Henry Kazina Quarry, north of Tyndall, 50°06'54"N, 96°40'24"W. *Hiscobeccus capax.*

MMMN I-2788, lower Selkirk Member, Red River Formation. First ridge from west, ditch outcrop south side of section road, about 2 miles north of Garson, 50°06'24"N, 96°41'19"W. *Hypsiptycha occidens.*

MMMN I-2833, lower Selkirk Member, Red River Formation. Northwest of rubble heap, Kazina Quarry, north of Tyndall, 50°06'54"N, 96°40'24"W. *Oepikina lata.*

MMMN I-2847, lower Selkirk Member, Red River Formation. Northwest of rubble heap, Henry Kazina Quarry, north of Tyndall, 50°06'54"N, 96°40'24"W. *Oepikina lata.*

MMMN I-2852, lower Selkirk Member, Red River Formation. Large block in northwest corner of Henry

Kazina Quarry, north of Tyndall, 50°06'54"N, 96°40'24". *Kjaerina hartae*.

MMMN I-2853, lower Selkirk Member, Red River Formation. Large block in northwest corner of Henry Kazina Quarry, north of Tyndall, 50°06'54"N, 96°40'24"W. *Nasutimena fluctuosa*.

MMMN I-2855, lower Selkirk Member, Red River Formation. Large block in northwest corner of Henry Kazina Quarry, north of Tyndall, 50°06'54"N, 96°40'24"W. *Tetraphalerella neglecta*.

MMMN I-2906, Cat Head Member, Red River Formation. North of Tyndall, Henry Kazina Quarry, east of rubble pile, 50°06'54"N, 96°40'24"W. *Hiscobeccus capax*.

MMMN I-3091-3104, Penitentiary Member, Stony Mountain Formation. East pit, City of Winnipeg Quarry, Stony Mountain. *Dinorthis occidentalis, Nasutimena fluctuosa, Rhynchotrema iowense, Hiscobeccus gigas*.

Plate 1

Figs. 1–22. *Dinorthis occidentalis* (Okulitch, 1943)

1–6. Hypotype, GSC 117745, dorsal, ventral, posterior, anterior, and lateral views of an articulated shell, and further enlarged posterior view showing strong cardinal process, open delthyrium and notothyrium, and well-developed ventral interarea; figs. 1–5, ×4; fig. 6, ×8.5; GSC loc. O-37130, Gunn Member, Stony Mountain Formation.

7–13. Hypotype, GSC 117746, 7–11, ventral, dorsal, posterior, anterior, and lateral views of an articulated shell, ×2.8; 12, local enlarged anterior view showing anterior commissure and some small holes on the radial ribs, ×6; 13, further enlarged posterior view showing well-differentiated cardinal process, open delthyrium and notothyrium, and high ventral interarea, ×6.2; same locality and horizon.

14–17. Hypotype, GSC 117747, ventral exterior, interior, posterior, and enlarged views of muscle field, showing well-impressed adductor and diductor scars, and strong hinge teeth; figs. 14–16, ×2; fig. 17, ×3; same locality and horizon.

18–22. Hypotype, GSC 117748, dorsal, posterior, anterior, ventral, and lateral views of an articulated shell, ×2.3; same locality and horizon.

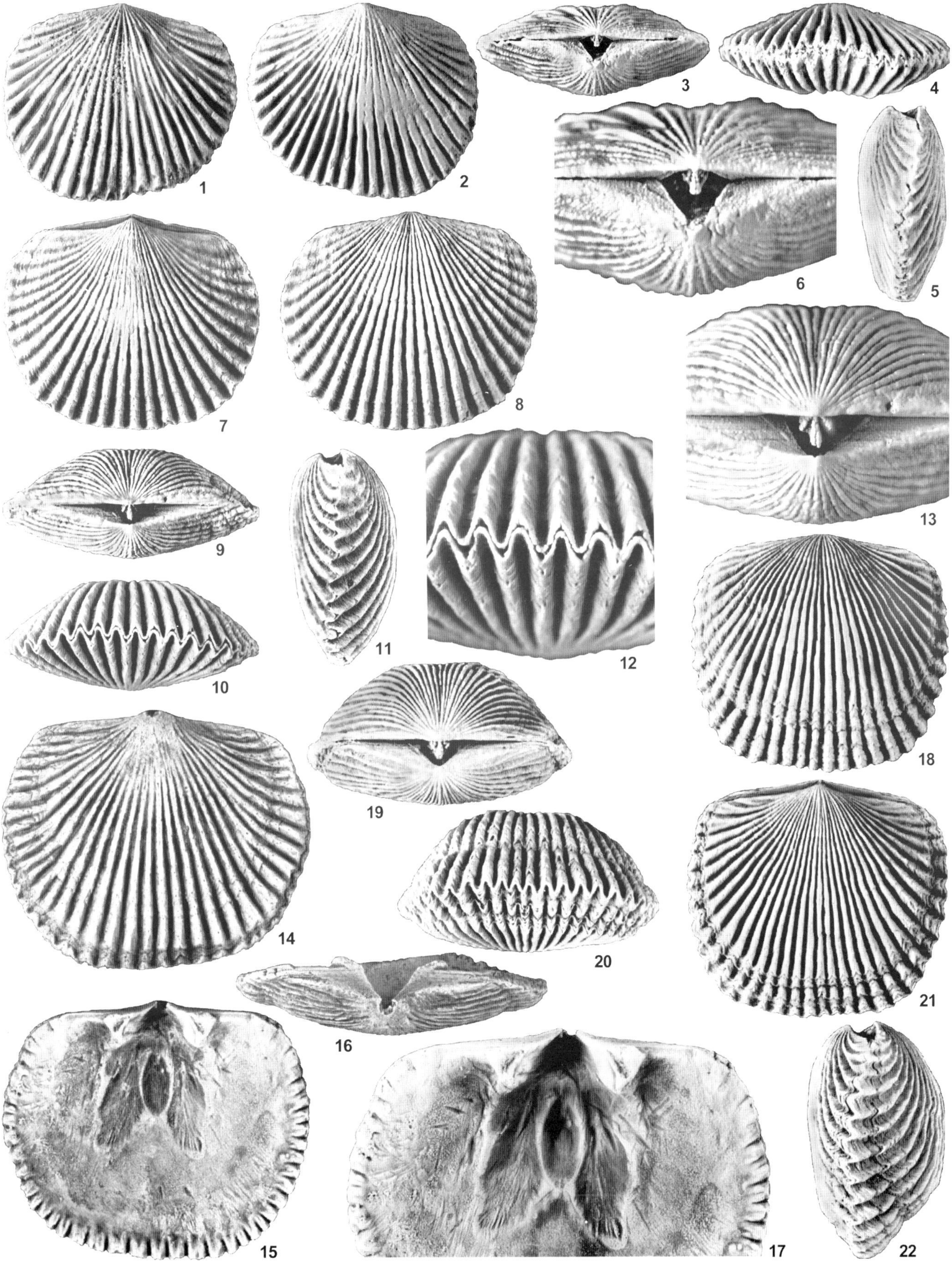
1
2
3
4
5
6
7
8
9
10
11
12
13
14
15
16
17
18
19
20
21
22

Plate 2

Figs. 1–13. *Dinorthis occidentalis* (Okulitch, 1943)

1, 2. Hypotype, MMMN I-3101, ventral and dorsal views of complete internal mould, ×2. East pit, City of Winnipeg Quarry, southern Manitoba; Penitentiary Member, Stony Mountain Formation.

3, 4. Hypotype, MMMN I-3100, ventral and dorsal views of complete internal mould, ×2. Same locality and horizon.

5, 6. Hypotype, MMMN I-3099, ventral and dorsal views of complete internal mould, ×2. Same locality and horizon.

7. Hypotype, GSC 108992, ventral interior showing subquadrate muscle field, with elongated oval and well-impressed adductor scars laterally enclosed by large, corrugated diductor scars, and well-developed palial markings, ×2. Stony Mountain; Gunn Member, Stony Mountain Formation.

8–10. Hypotype, GSC 108994, dorsal interior (10, ×2) and local enlargement showing cardinalia, sockets, brachiophore fossettes muscle field (8, ×3), and single ridge-like cardinal process myophore (9, ×4). Same locality and horizon.

11. Hypotype, GSC 108993, ventral interior showing subquadrate muscle field, with elongated oval and deeply impressed adductor scars laterally enclosed by large, corrugated diductor scars, ×2. Same locality and horizon.

12, 13. Hypotype, MMMN I-3098, ventral and dorsal views of complete internal mould, ×2. East pit, City of Winnipeg Quarry, southern Manitoba; Penitentiary Member, Stony Mountain Formation.

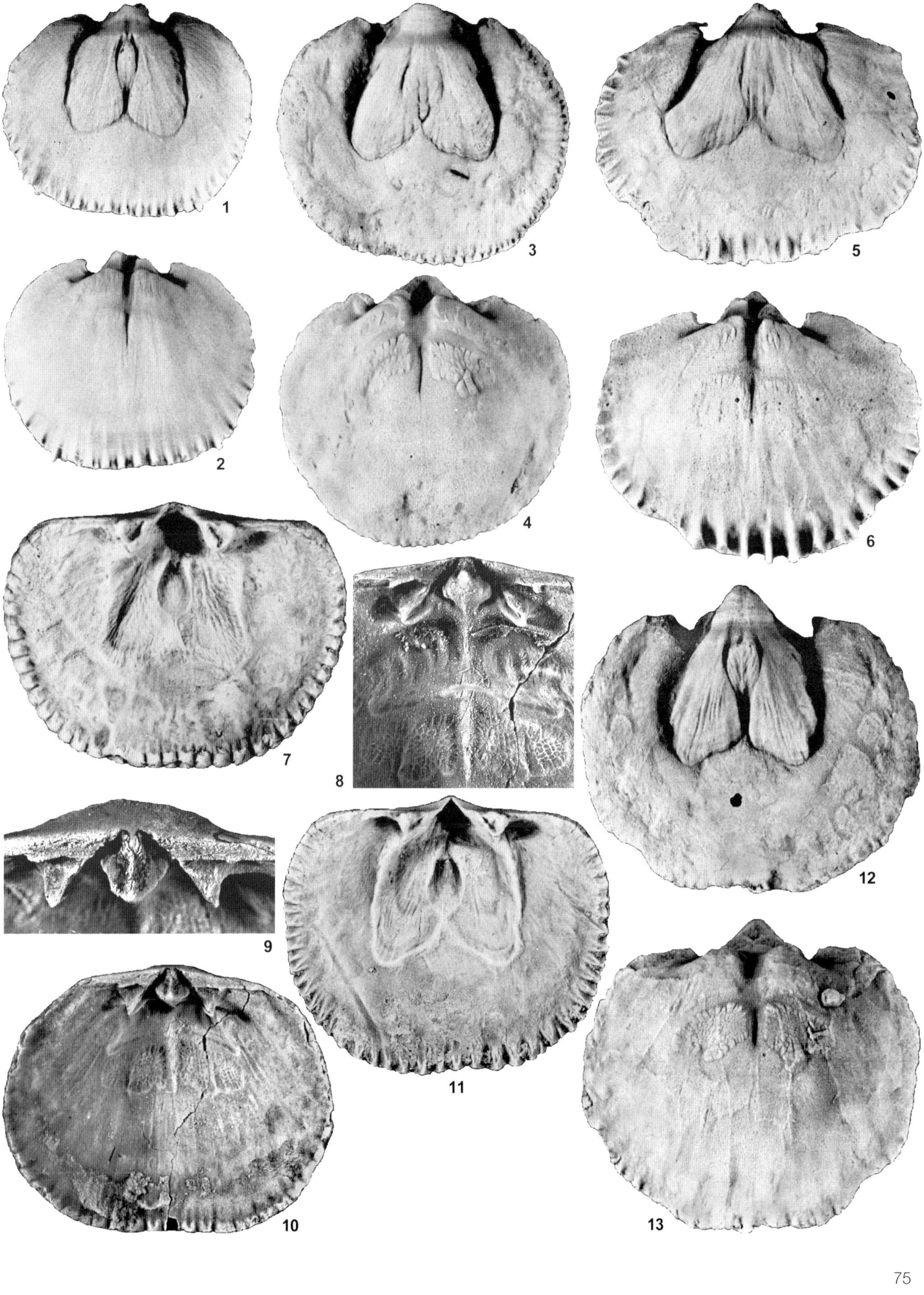
1
2
3
4
5
6
7
8
9
10
11
12
13

Plate 3

Figs. 1–5. *Dinorthis occidentalis* (Okulitch, 1943)

Hypotype, GSC 117749, ventral, dorsal, posterior, anterior, and lateral views of an articulated shell, ×2; GSC loc. O-37130, Gunn Member, Stony Mountain Formation.

Figs. 6–20. *Gnamptorhynchos manitobensis* Jin and Zhan, 2000

6–10. Paratype, GSC 117739, dorsal, ventral, lateral, posterior, and anterior views of an articulated shell, ×3; Garson Quarry, Selkirk Member, Red River Formation.

11–15. Paratype, GSC 117740, dorsal, posterior, anterior, lateral, and ventral views of an articulated shell, ×2.5; same locality and horizon.

16–20. Holotype, GSC 117741, dorsal, ventral, lateral, posterior, and anterior views of an articulated shell, ×2.5; same locality and horizon.

1
2
3
4
5
6
7
8
9
10
11
12
13
14
15
16
17
18
19
20

Plate 4

Figs. 1–20. *Diceromyonia storeya* (Okulitch, 1943)

1–5. Hypotype, GSC 117750, dorsal, ventral, lateral, posterior, and anterior views of an articulated shell, ×4; loc. O-37130, Gunn Member, Stony Mountain Formation.

6, 7. Hypotype, GSC 117751, dorsal interior and exterior views, ×3; same locality and horizon.

8, 9. Hypotype, GSC 117752, dorsal interior and exterior views, ×4; same locality and horizon.

10–12. Hypotype, GSC 117753, dorsal interior, exterior, and posterior enlarged views, showing well-differentiated cardinal process, strong brachiophores and thick myophragm of the muscle field; figs. 10, 11, ×4; fig. 12, ×5; same locality and horizon.

13–17. Hypotype, GSC 117754, dorsal, ventral, posterior, anterior, and lateral views of an articulated shell, ×4; same locality and horizon.

18–20. Hypotype, GSC 117755, ventral exterior, interior, and enlarged views, showing hinge teeth, accessory teeth, dental plates, and deeply impressed diductor and adjustor scars; figs. 18, 19, ×3; fig. 20, ×4; same locality and horizon.

Figs. 21, 22. *Thaerodonta clarksvillensis* (Foerste, 1912)

21. GSC 117756, ventral internal mould, ×5; loc. 55-60, Penitentiary Member, Stony Mountain Formation.

22. GSC 117757, dorsal internal mould, ×5; same locality and horizon.

1
2
3
4
5
6
7
8
9
10
11
12
13
14
15
16
17
18
19
20
21
22

Plate 5

Figs. 1–5. *Diceromyonia storeya* (Okulitch, 1943)

Holotype, GSC 2043, dorsal, ventral, posterior, anterior, and lateral views of an articulated shell, ×4; City of Winnipeg Quarry, Gunn Member, Stony Mountain Formation.

Figs. 6–8. *Thaerodonta clarksvillensis* (Foerste, 1912)

6. GSC 117756, enlarged view of ventral internal mould, showing well-developed fossettes corresponding to dorsal denticles, ×7; loc. 55-60, Penitentiary Member, Stony Mountain Formation.

7, 8. GSC 117758, dorsal internal and external moulds, showing strong undercut cardinal process, long side septa, denticles, and weak shell ornamentation, ×7; same locality and horizon.

Figs. 9–15. *Strophomena vetusta* James, 1874

9, 10. GSC 117759, dorsal interior and local enlargement showing cardinalia, notothyrial platform, clearly impressed muscle field and strong transmuscle ridges; fig. 9, ×2.5; fig. 10, ×4.7; loc. O-37130, Gunn Member, Stony Mountain Formation.

11–13. GSC 117760, ventral interior, posterior view, and local enlargement showing pseudodeltidium, muscle field and its bounding ridges; figs. 11, 12, ×2.5; fig. 13, ×3.9; same locality and horizon.

14, 15. GSC 117761, dorsal interior and local enlargement showing large, triangular cardinal process lobes, sockets, inner and outer socket ridges, notothyrial platform, deep muscle field, transmuscle ridges, and densely pustules on both sides of the shell internal surface; fig. 14, ×2.5; fig. 15, ×7; same locality and horizon.

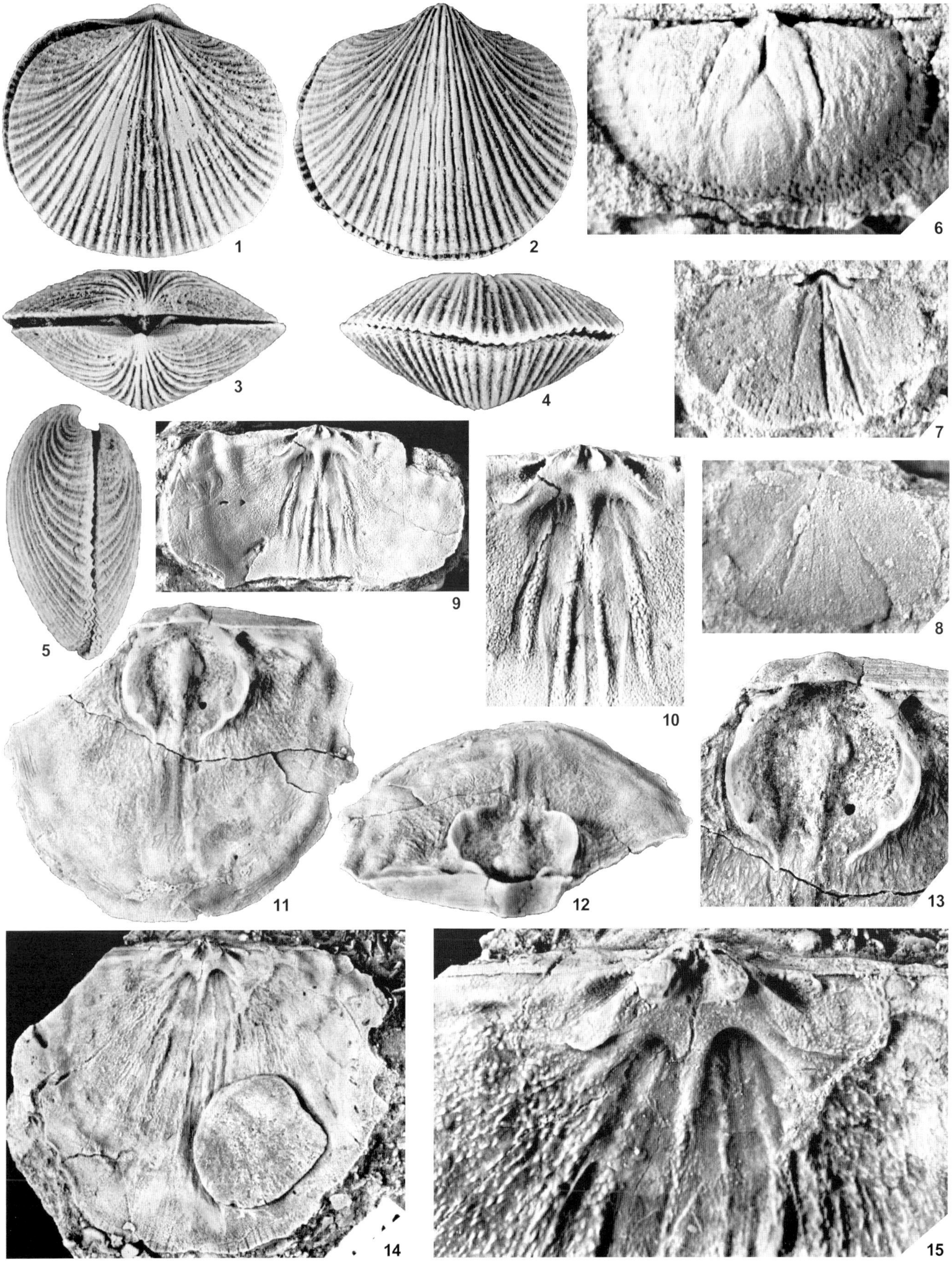
1
2
6
3
4
7
5
9
10
8
11
12
13
14
15

Plate 6

Figs. 1, 2. *Strophomena vetusta* James, 1874

GSC 117762, ventral and dorsal views of a broken articulated shell, ×2.5; loc. O-37130, Gunn Member, Stony Mountain Formation.

Figs. 3–8. *Strophomena planumbona* (Hall, 1847)

GSC 117763, ventral, dorsal, lateral, posterior, anterior, and local enlargement showing its ornamentation of parvicostellae and fine concentric growth lines; figs. 3–7, ×3; fig. 8, ×6.4; loc. O-37130, Gunn Member, Stony Mountain Formation.

Figs. 9–19. *Nasutimena fluctuosa* (Billings, 1860)

9–14. GSC 117764, ventral, lateral, dorsal, posterior, anterior, and local enlargement showing its ornamentation of parvicostellae and concentric rugae; figs. 9–13, ×3; fig. 14, ×5.5; same locality and horizon.

15, 16. GSC 117765, posterior view of an articulated shell and its local enlargement showing high ventral interarea and well-developed pseudodeltidium, fig. 15, ×3; fig. 16, ×8; same locality and horizon.

17–19. GSC 117766, ventral interior and posterior views, and its local enlargement showing deep ventral muscle field, bounding ridges, and myophragm; figs. 17, 18, ×3; figs. 19, ×6; same locality and horizon.

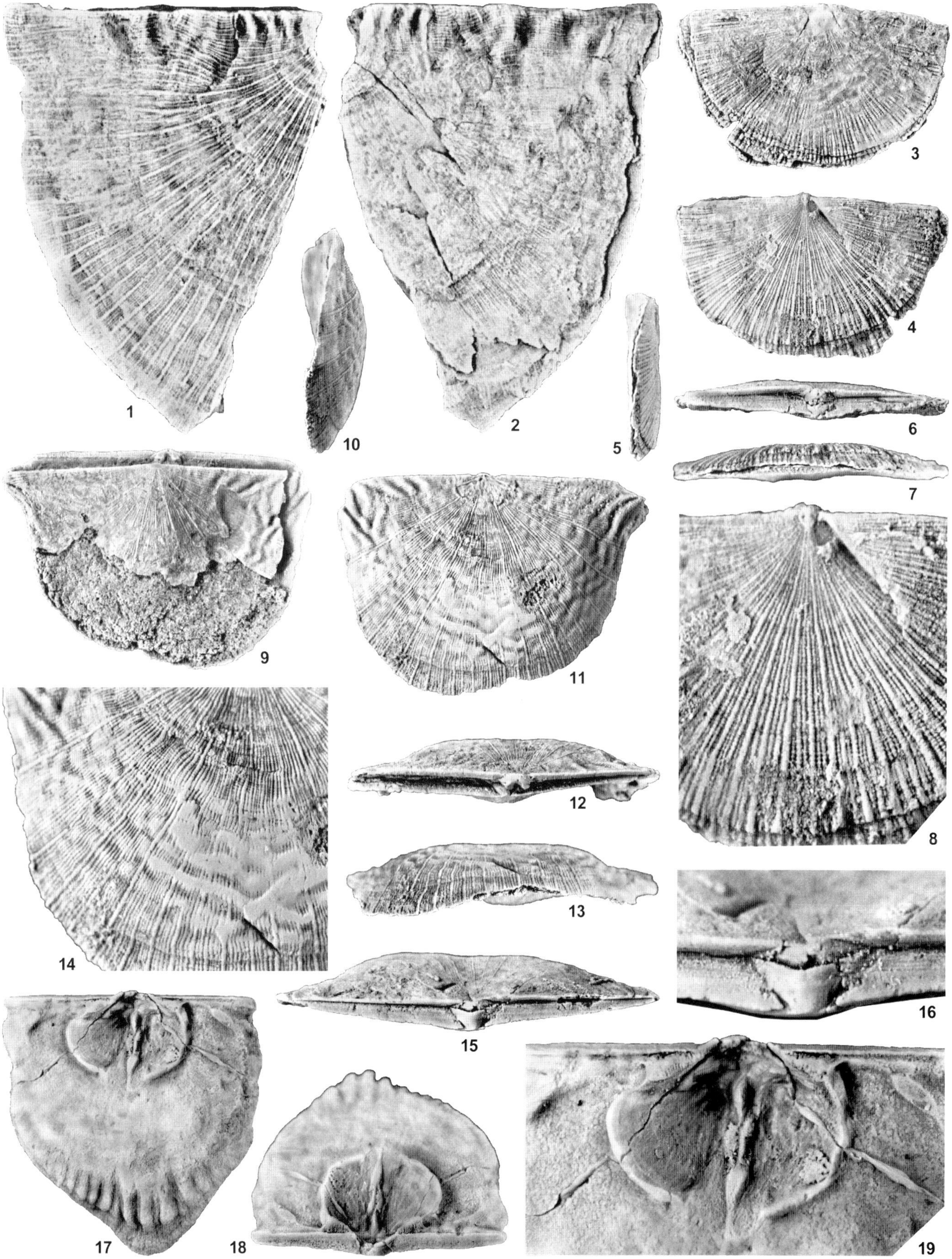
1
10
2
5
3
4
6
7
9
11
8
14
12
13
15
16
17
18
19

Plate 7

Figs. 1–4. *Strophomena planumbona* (Hall, 1847)

GSC 117765, dorsal, ventral, lateral, and anterior views of an articulated shell, ×3; GSC loc. O-37130, Gunn Member, Stony Mountain Formation.

Figs. 5–12. *Nasutimena fluctuosa* (Billings, 1860)

5–9. MMMN I-2153a, lateral, posterior, ventral, anterior, and dorsal views of an articulated shell, ×2.5; Garson Quarry, Selkirk Member, Red River Formation.

10–12. MMMN I-2153b, dorsal lateral, posterior, and exterior views of an articulated shell; figs. 10, 12, ×2.5; fig. 11, ×3; same locality and horizon.

Figs. 13, 14. *Nasutimena undulosa* (Roy, 1941)

Hypotype, GSC 109025, posterior and lateral views of a broken, articulated shell, ×1.5; Garson Quarry, Selkirk Member, Red River Formation.

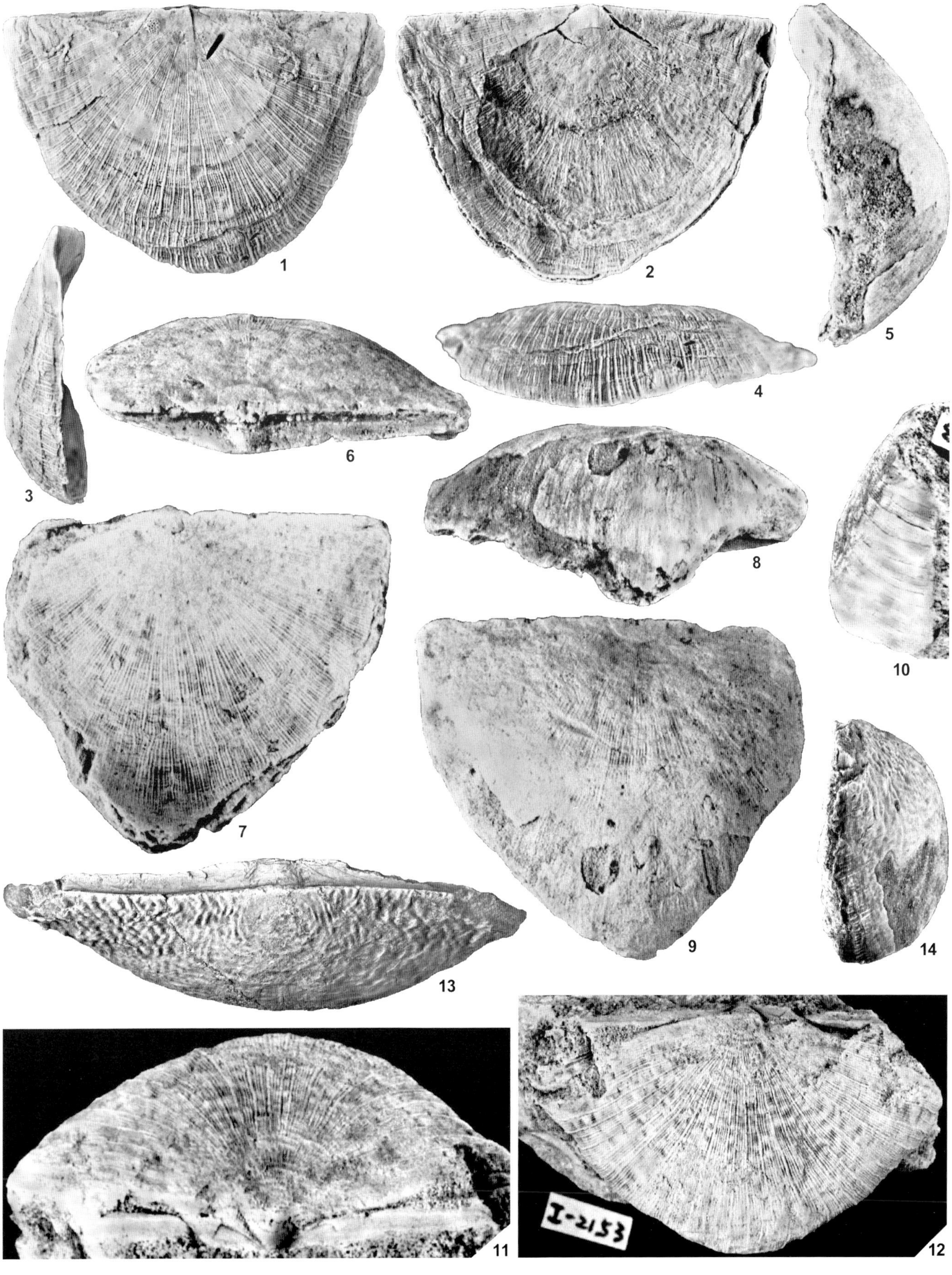
1
2
3
4
5
6
7
8
9
10
11
12
13
14
I-2153

Plate 8

Figs. 1–4. *Nasutimena undulosa* (Roy, 1941)

1, 2. Hypotype, GSC 109025, dorsal and ventral views of the same shell as Pl. 7, figs. 13, 14, ×1.5; Garson Quarry, Selkirk Member, Red River Formation.

3, 4. Hypotype, GSC 109024, dorsal and visceral disc views of a dorsal valve with strong criss-cross corrugations covering disc and trail, ×1.5; Garson Quarry, Selkirk Member, Red River Formation.

Figs. 5–8. *Holtedahlina paraprostrata* n. sp.

5, 6. Paratype, MMMN I-2543a, ventral internal and external moulds, ×4.3; west bank of Red River about 300 m south of Parkes Creek, Fort Garry Member, Red River Formation.

7. Holotype, MMMN I-2543b, dorsal internal mould, ×6.3; same locality and horizon.

8. Paratype, MMMN I-2543c, dorsal internal mould, ×5.5; same locality and horizon.

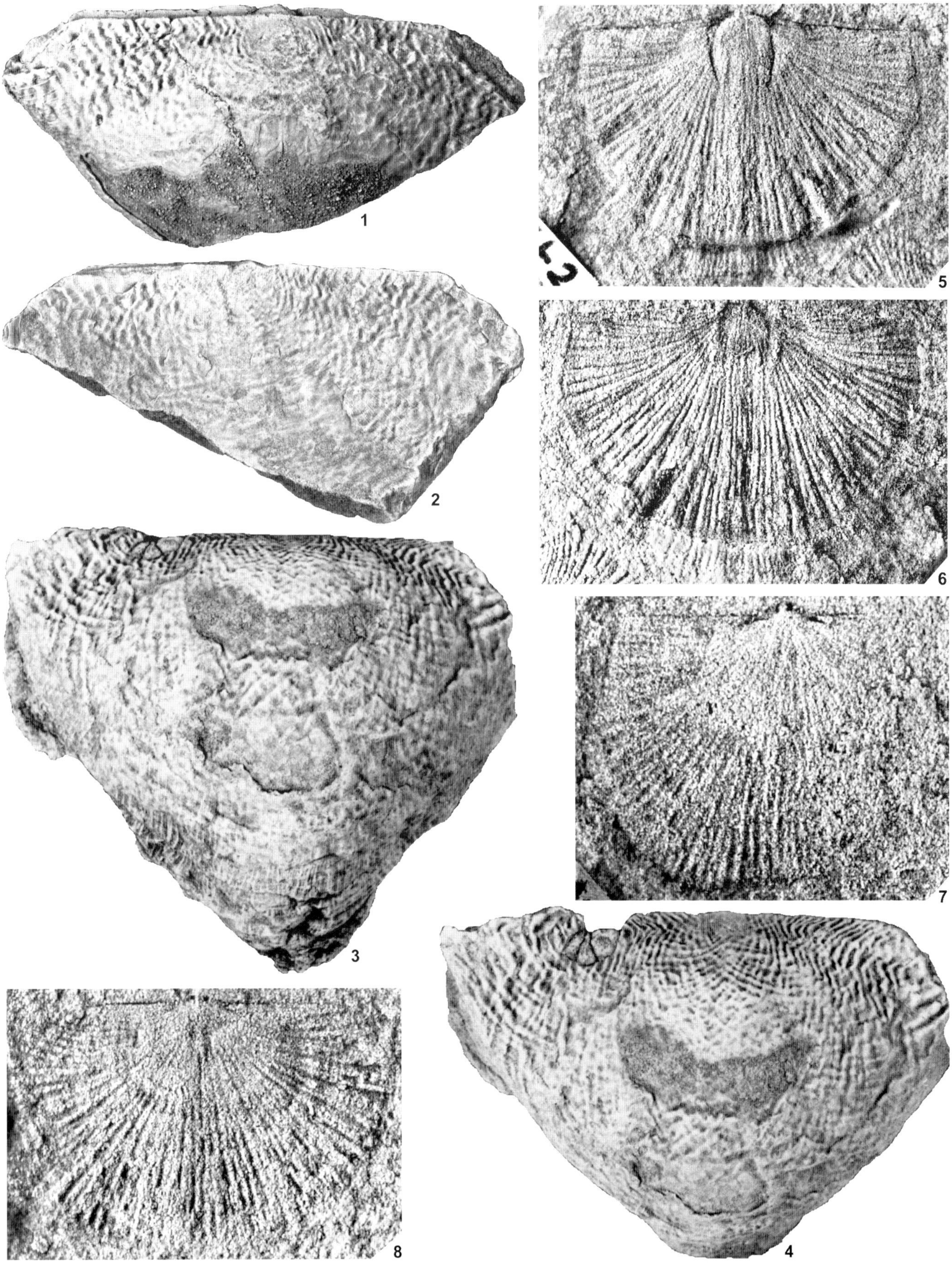

Plate 9

Figs. 1, 2, 8, 9. *Tetraphalerella churchillensis* Jin, Caldwell, and Norford, 1997

1, 2. Hypotype, GSC 109046, dorsal exterior (2) showing postero–lateral rugae, and local enlargement (1) showing equal-sized costellae and coarse pseudopunctae; fig. 1, ×3.3; fig. 2, ×1.5; Garson Quarry, Selkirk Member, Red River Formation.

8. Hypotype, MMMN I-185, ventral exterior, ×1.5; same locality and horizon.

9. Hypotype, MMMN I-1678, ventral exterior, ×1.5; same locality and horizon.

Figs. 3–7. *Tetraphalerella neglecta* (James, 1881)

3–5. MMMN I-298, two dorsal internal moulds and their enlargement, respectively; fig. 3, ×1; figs. 4, 5, ×2; Cat Head, Cat Head Member, Red River Formation.

6. GSC 117767, ventral external mould, ×2.5; Birch Island, Lake Winnipeg, Cat Head Member, Red River Formation.

7. GSC 117768, dorsal external mould, ×2.5; same locality and horizon.

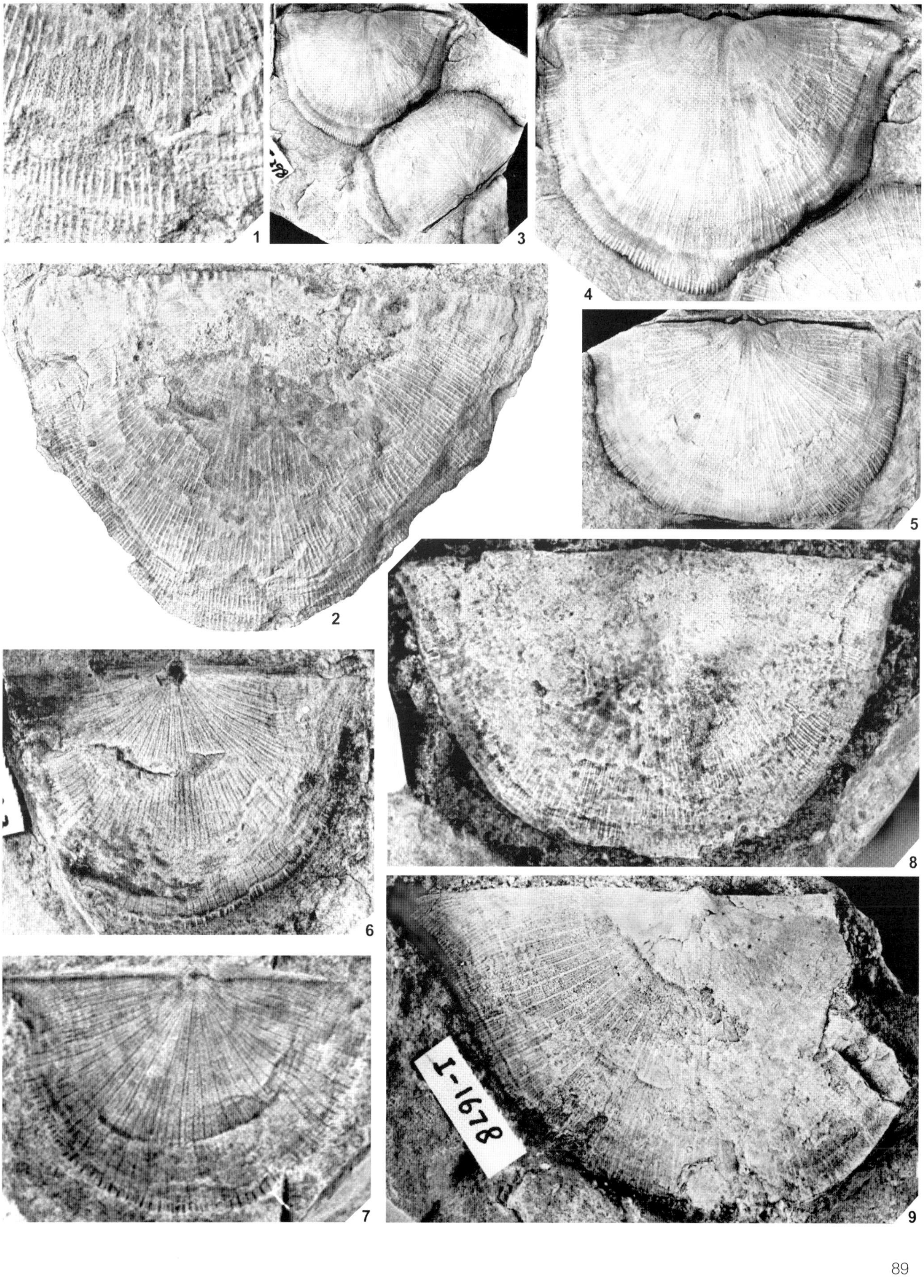
1
3
4
5
2
8
6
7
9
I-1678

Plate 10

Figs. 1–13. *Oepikina lata* (Whiteaves, 1896)

1–5. Hypotype, MMMN I-2151, dorsal, ventral, lateral, posterior, and anterior views of an articulated shell, ×1.5; Garson Quarry, Selkirk Member, Red River Formation.

6–10. Hypotype, MMMN I-1891, dorsal, lateral, posterior, ventral, and anterior views of an articulated shell, ×1.5; same locality and horizon.

11–13. Hypotype, MMMN I-2847, ventral exterior, posterior, and lateral views showing disc and geniculation, ×1.5; Kazina Quarry, north of Tyndall, Selkirk Member, Red River Formation.

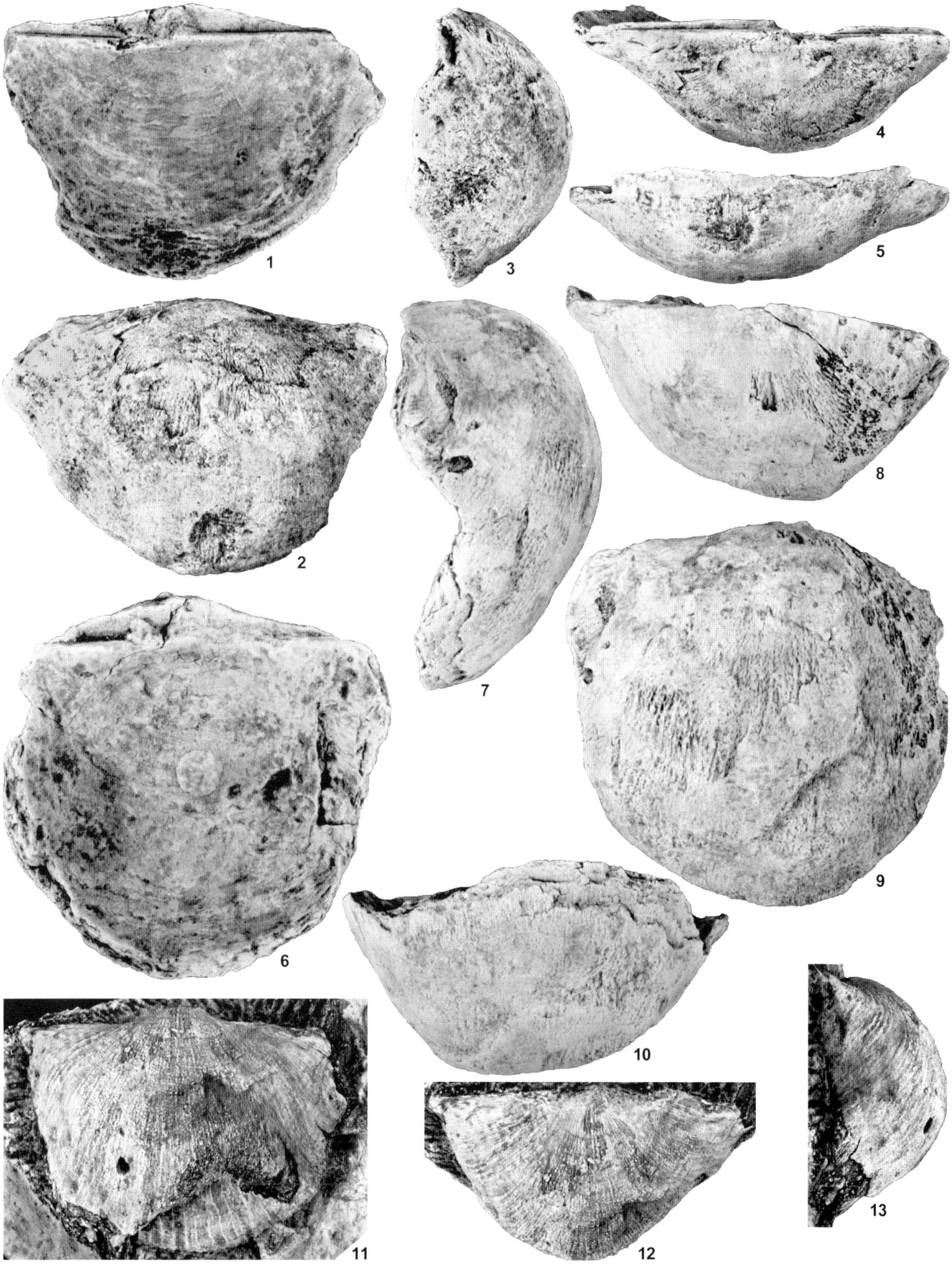

Plate 11

Figs. 1–10. *Oepikina lata* (Whiteaves, 1896)

1–3. Holotype, GSC 4391, dorsal, ventral, and lateral views of a partially dolomitized shell, Lower Fort Garry, Selkirk Member, Red River Formation.

4, 5. Hypotype, RJE 1–128, ventral and lateral views of a broken shell, ×1.5, ×1; Garson Quarry, Selkirk Member, Red River Formation.

6, 7. Hypotype, GSC 109086, interior of a dorsal valve with well-preserved transmuscle septa and a slightly damaged, bilobed cardinal process; fig. 6, ×1.5; fig. 7, ×2.2; same locality and horizon.

8–10. Hypotype, MMMN I-2150, lateral, posterior, and ventral views of a shell embedded in matrix, ×1.5; same locality and horizon.

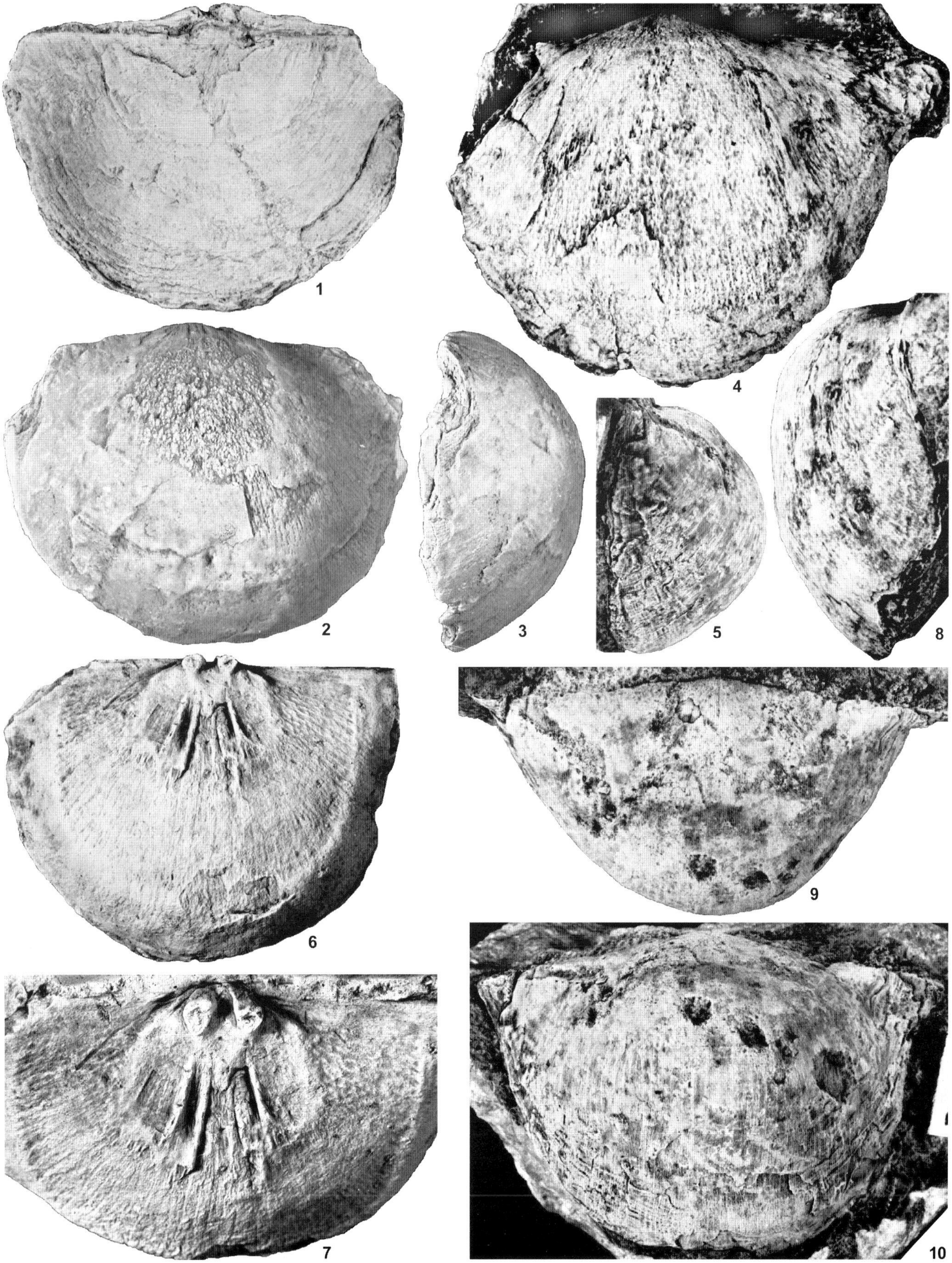
1
2
3
4
5
6
7
8
9
10

Plate 12

Figs. 1–21. *Oepikina limbrata* Wang, 1949

1–6. Hypotype, GSC 117769, dorsal, ventral, lateral, posterior, and anterior views of an articulated shell, and its posterior enlargement showing strong, postero-ventrally projecting cardinal process, arched pseudodeltidium, and well-developed ventral and dorsal interareas; figs. 1–5, ×2.5; fig. 6, ×5; GSC loc. O-37130, Gunn Member, Stony Mountain Formation.

7, 8. Hypotype, GSC 117770, ventral interior and local enlargement showing its arched pseudodeltidium, small teeth, strongly elevated adductor scars, and clearly impressed circular diductor scars; fig. 7, ×2.5; fig. 8, ×4.6; same locality and horizon.

9. Hypotype, GSC 117771, dorsal interior showing highly projecting cardinal process, ridge-like socket ridges, variously developed myophragm and transmuscle ridges, and clear peripheral rim; ×2.5; same locality and horizon.

10. Hypotype, GSC 117772, dorsal interior showing nearly circular visceral area, ×2.5; same locality and horizon.

11–15. Hypotype, GSC 117773, posterior, anterior, lateral, dorsal, and ventral views of an articulated shell, ×2.5; same locality and horizon.

16–21. Hypotype, GSC 117774, 16–18, 20, 21, 19, dorsal, ventral, lateral, posterior, and anterior views of an articulated shell, and its posterior enlargement showing well-developed pseudodeltidium and chilidium, small foramen, and high ventral and dorsal interareas; figs. 16–18, 20, 21, ×2.5; fig. 19, ×9.3; same locality and horizon.

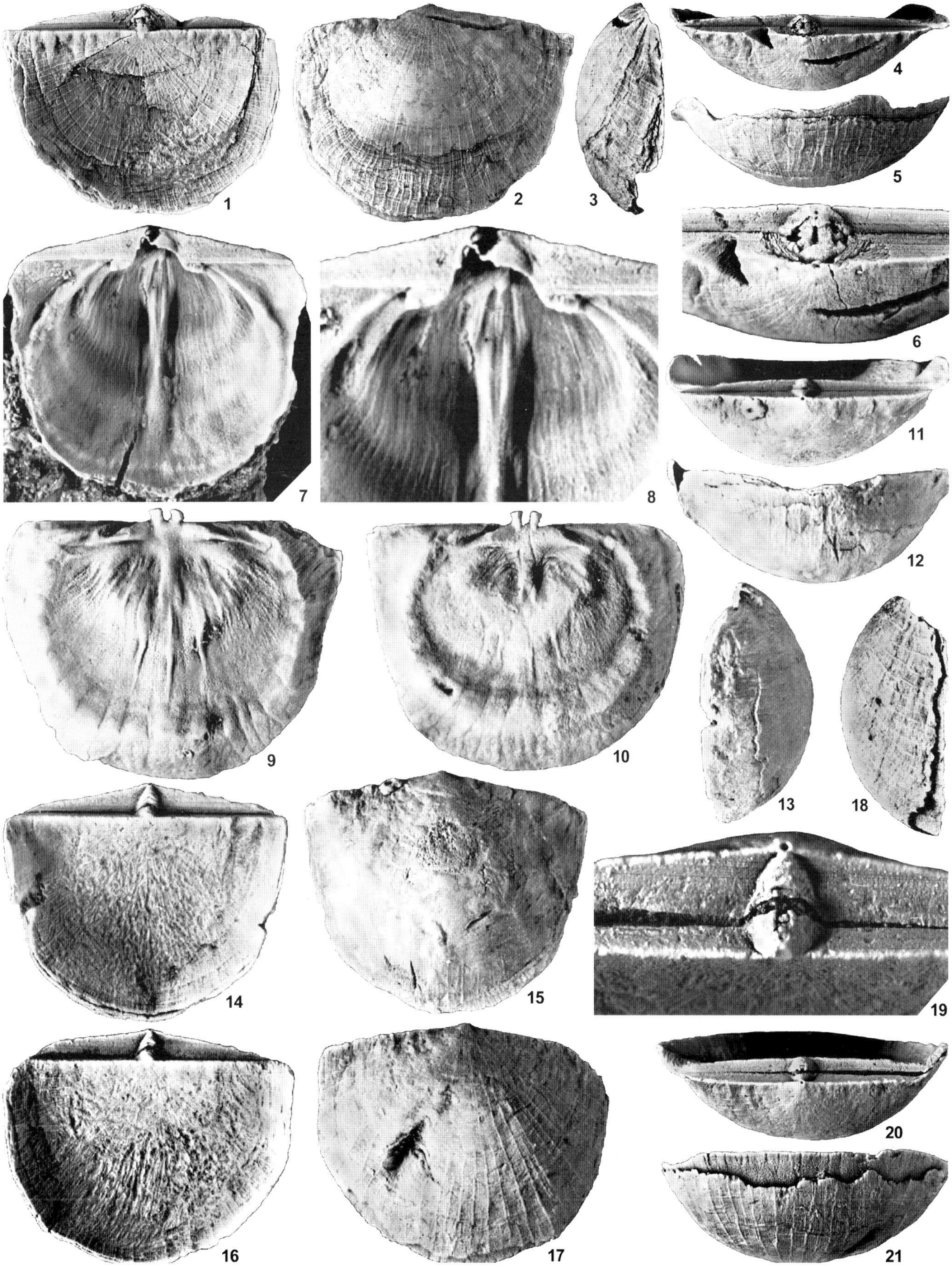
1
2
3
4
5
6
7
8
9
10
11
12
13
14
15
16
17
18
19
20
21

Plate 13

Figs. 1–14. *Kjaerina hartae* Jin, Caldwell, and Norford, 1995

1–4. Hypotype, UM I-94a, ventral, dorsal, anterior, and posterior views of an articulated shell, ×1.5; Garson Quarry, Selkirk Member, Red River Formation.

5, 13, 14. Hypotype, UM 5161, lateral, posterior, and ventral views of a ventral exterior; ×1.5; same locality and horizon.

6–8. Hypotype, UM I-94, lateral, posterior, and ventral views of a ventral exterior, showing long geniculation, transverse disc, and well-developed concentric rugae, ×1.5; same locality and horizon.

9–12. Holotype, GSC 109077, lateral, dorsal, ventral, and anterior views of a strongly transverse shell with sharp geniculation, ×2; same locality and horizon.

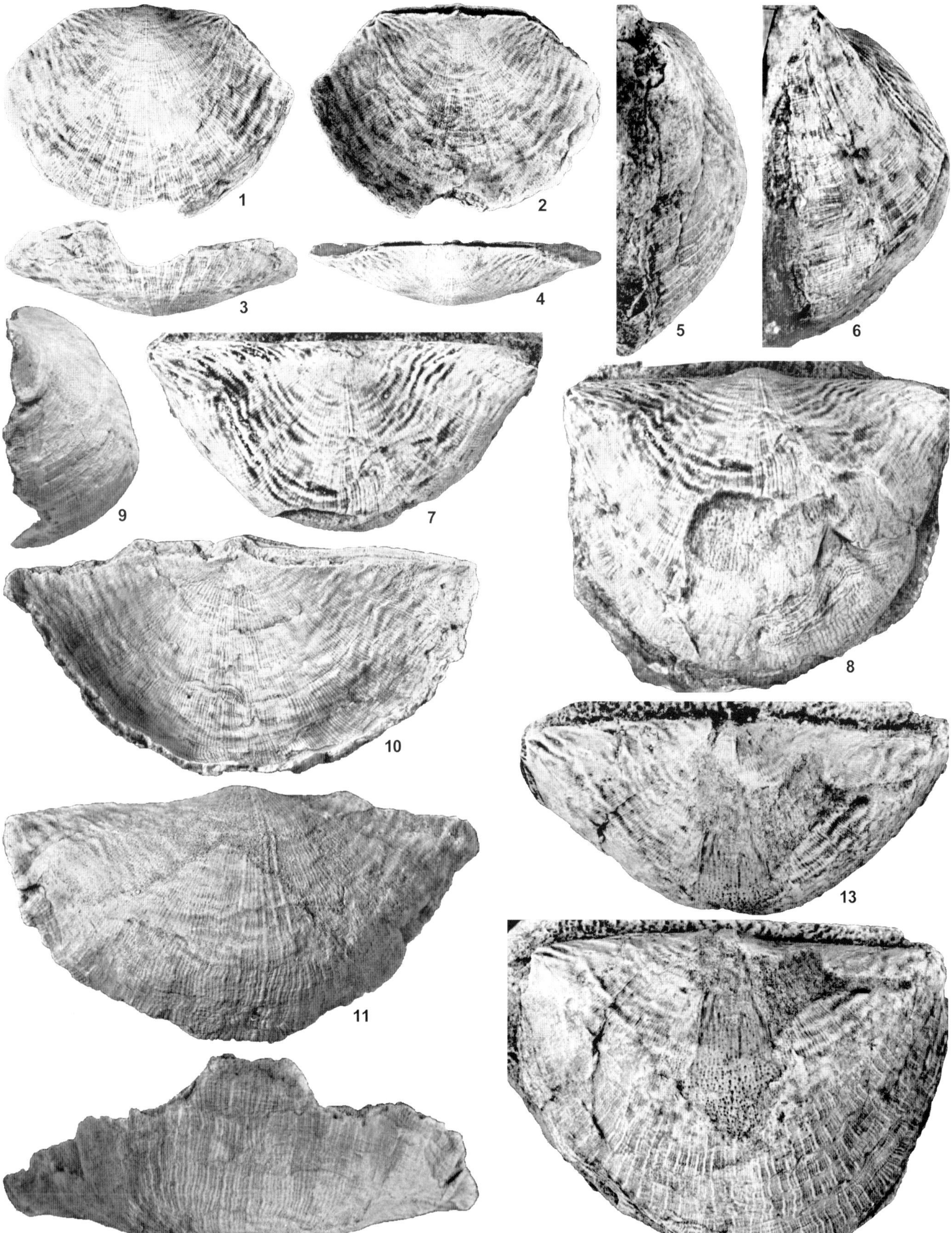
1
2
3
4
5
6
7
8
9
10
11
12
13
14

Plate 14

Figs. 1–13. *Megamyonia nitens* (Billings, 1860)

1–6. Hypotype, GSC 117775, dorsal, ventral, lateral, posterior, and anterior views of an articulated shell, and its posterior enlargement showing high ventral interarea, low dorsal interarea, and poorly developed pseudodeltidium and chilidium; figs. 1–5, ×3; fig. 6, ×7.2; GSC loc. O-37130, Gunn Member, Stony Mountain Formation.

7–12. Hypotype, GSC 117776, dorsal, ventral, anterior, posterior, and lateral views of an articulated shell, and its posterior enlargement showing high ventral interarea, pseudodeltidium, and chilidium; figs. 7–11, ×3; fig. 12, ×9.8; same locality and horizon.

13. Hypotype, GSC 117777, ventral view of an articulated shell, ×3; same locality and horizon.

Figs. 14–18. *Parastrophinella cirrita* n. sp.

Holotype, GSC 117778, dorsal, ventral, lateral, posterior, and anterior views of an articulated shell, showing its dorsibiconvex lateral profile and strong plications on the shell anterior one-third, ×5; Garson Quarry, Selkirk Member, Red River Formation.

Figs. 19–22. *Rhynchotrema increbescens* (Hall, 1847)

MMMN 970605, dorsal, ventral, lateral, and posterior views of a complete internal mould, ×5; City Quarry, Stony Mountain, base of Penitentiary Member, Stony Mountain Formation.

1
2
3
4
5
6
7
8
9
10
11
12
13
14
15
16
17
18
19
20
21
22

Plate 15

Figs. 1–19. *Rhynchotrema iowense* Wang, 1949

1–4. Hypotype, MMMN I-3091, dorsal, ventral, lateral, and posterior views of a complete internal mould, ×4; City Quarry, Stony Mountain, Penitentiary Member, Stony Mountain Formation.

5–9. Hypotype, MMMN I-3092, posterior, ventral, dorsal, and lateral views of a complete internal mould, and its local enlargement showing blade-like cardinal process, septalium, dorsal median septum, and subparallel dental plates; figs. 5, 7–9, ×4; fig. 6, ×9.3; same locality and horizon.

10–14. Hypotype, GSC 117779, dorsal, ventral, lateral, posterior, and anterior views of an articulated shell, ×3; Garson Quarry, Selkirk Member, Red River Formation.

15–19. Hypotype, GSC 117780, dorsal, ventral, lateral, posterior, and anterior views of an articulated shell, ×4; same locality and horizon.

Figs. 20–24. *Hypsiptycha anticostiensis* (Billings, 1862)

Hypotype, GSC 117781, dorsal, lateral, posterior, anterior, and ventral views of an articulated shell, ×3; Garson Quarry, Selkirk Member, Red River Formation.

1
2
3
6
4
5
12
7
10
11
8
9
15
13
14
17
16
21
18
19
20
22
23
24

Plate 16

Figs. 1–5. *Lepidocyclus laddi* Wang, 1949

Hypotype, GSC 117782, dorsal, ventral, lateral, posterior, and anterior views of an articulated shell, ×2; Garson Quarry, Selkirk Member, Red River Formation.

Figs. 6–17. *Hypsiptycha anticostiensis* (Billings, 1862)

6–10. Hypotype, GSC 117783, dorsal, ventral, lateral, anterior, and posterior views of an articulated shell, ×3; same locality and horizon.

11–17. Hypotype, GSC 117784, dorsal, ventral, posterior, anterior, and lateral views of an articulated shell, and local enlargements (16, 17) showing well-preserved growth lamellae, high ventral beak, wide and rounded palintrope; figs. 11–15, ×3; fig. 16, ×7; fig. 17, ×6.2; same locality and horizon.

1
2
3
4
11
6
8
5
10
12
7
9
14
13
16
15
17

Plate 17

Figs. 1–19. *Hypsiptycha occidens* (Wilson, 1926)

1–6. Hypotype, GSC 117785, dorsal, ventral, lateral, posterior, and anterior views of an articulated shell, and local enlargement showing erect ventral beak and well-developed deltidial plates; figs. 1–5, ×5; fig. 6, ×9.2; loc. O-37130, Gunn Member, Stony Mountain Formation.

7–12. Hypotype, GSC 117786, dorsal, ventral, lateral, posterior, and anterior views of an articulated shell, and posterior enlargement showing high ventral umbo, well-developed and medially connected deltidial plates, and foramen; figs. 7–11, ×5; fig. 12, ×12.5; same locality and horizon.

13–19. Hypotype, GSC 117787, dorsal, ventral, lateral, posterior, and anterior views of an articulated shell, and local enlargements (18, 19) showing medially connected deltidial plates, foramen, wide palintrope, and well-preserved growth lamellae; figs. 13–17, ×5; fig. 18, ×10.9; fig. 19, ×11.3; same locality and horizon.

1
2
3
4
5
6
7
8
9
10
11
12
13
14
15
16
17
18
19

Plate 18

Figs. 1–7. *Hypsiptycha occidens* (Wilson, 1926)

Hypotype, GSC 117788, dorsal, ventral, lateral, posterior, and anterior views of an articulated shell, and local enlargements (6, 7) showing deltidial plates, foramen, and strong growth lamellae; figs. 1–5, ×3.5; fig. 6, ×5.7; fig. 7, ×9.2; loc. O-37130, Gunn Member, Stony Mountain Formation.

Figs. 8–17. *Hiscobeccus capax* (Conrad, 1842)

8–12. Hypotype, GSC 117789, dorsal, posterior, ventral, lateral, and anterior views of an articulated shell, ×3; Garson Quarry, Selkirk Member, Red River Formation.

13–17. Hypotype, MMMN I-2906, dorsal, posterior, lateral, ventral, and anterior views of an articulated shell, ×3; Kazina Quarry, north of Tyndall, Cat Head Member, Red River Formation.

1
2
3
4
5
6
7
8
9
10
11
12
13
14
15
16
17

Plate 19

Figs. 1–11. *Hiscobeccus gigas* (Wang, 1949)

1–5. Hypotype, GSC 117790, dorsal, ventral, lateral, posterior, and anterior views of an articulated shell, ×2; loc. O-37130, Gunn Member, Stony Mountain Formation.

6–10. Hypotype, GSC 117791, dorsal, posterior, anterior, ventral, and lateral views of an articulated shell, ×2.5; same locality and horizon.

11. Hypotype, GSC 117792, one of the serial sections (SM 4-9, 2.0 mm away from the shell posterior end) showing high ridge-like cardinal process in the notothyrial cavity, ×28.6; same locality and horizon.

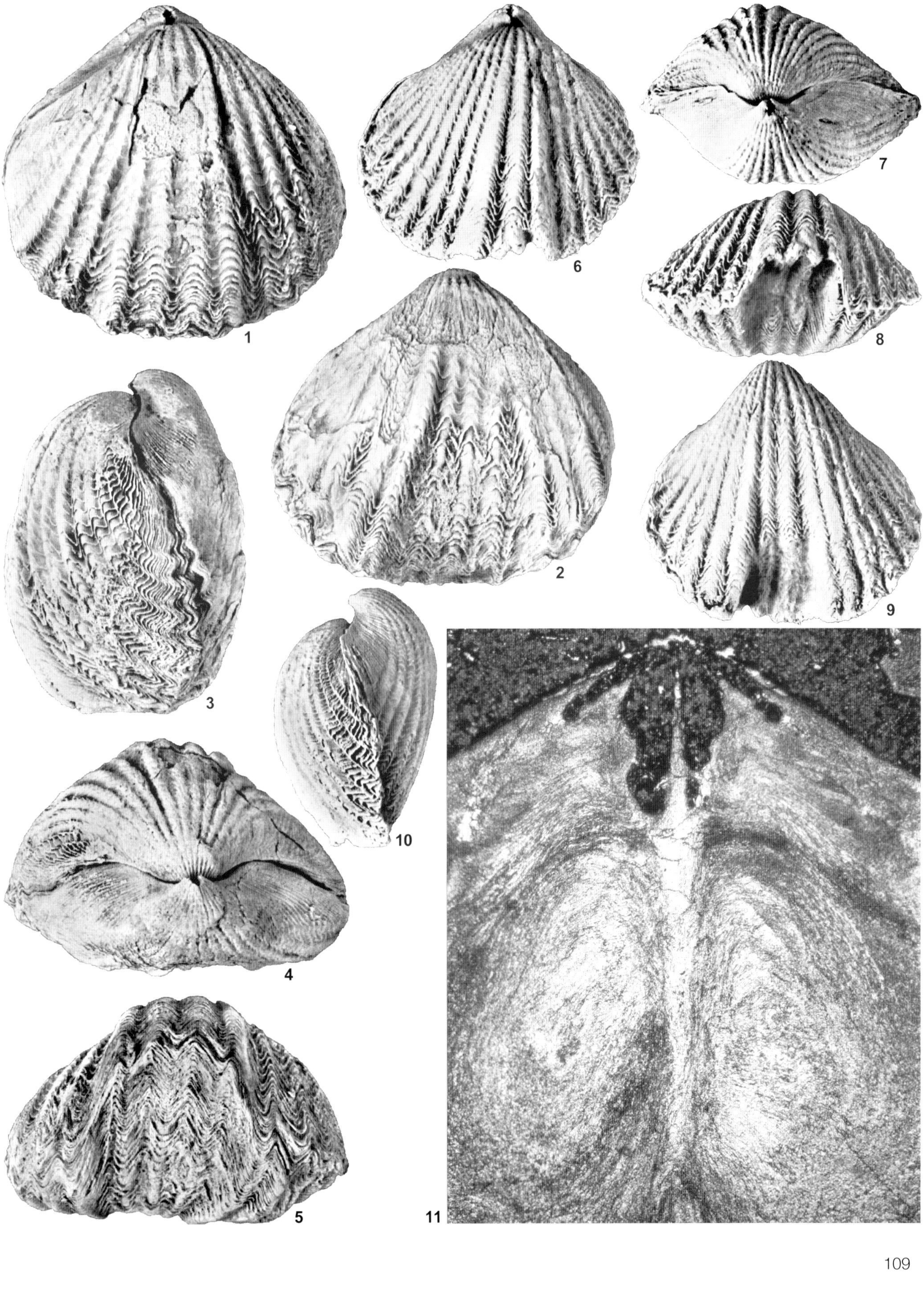
1
2
3
4
5
6
7
8
9
10
11

Plate 20

Figs. 1–7. *Nasutimena fluctuosa* (Billings, 1860)

1, 2. Hypotype, MMMN I-3097, ventral and dorsal views of complete internal mould, ×2.5. Penitentiary Member, Stony Mountain Formation; East pit, City of Winnipeg Quarry.

3, 4. Hypotype, MMMN I-3096, ventral and dorsal views of a complete internal mould, ×2.5; same locality and horizon.

5. Hypotype, MMMN I-3095, dorsal internal mould showing cardinalia, muscle field, and side septa, ×2.5; same locality and horizon.

6. Hypotype, MMMN I-3094, dorsal internal mould, ×2.5; same locality and horizon.

7. Hypotype, MMMN I-3093, dorsal internal mould showing cardinalia, median ridge, and side septa, ×2.5. Same locality and horizon.

Figs. 8–13. *Hiscobeccus gigas* (Wang, 1949)

8, 9. Hypotype, MMMN I-3104, ventral and dorsal views of complete internal mould showing large and well-impressed ventral muscle field, palial markings, and long dorsal median septum, ×2; same locality and horizon.

10, 11. Hypotype, MMMN I-3102, ventral and dorsal views of complete internal mould, ×2; same locality and horizon.

12, 13. Hypotype, MMMN I-3103, ventral and dorsal views of complete internal mould showing large ventral muscle field with elongated oval adductor scars completely enclosed by striated diductor scars, cardinal process continuous with strong and high median septum, ×2; same locality and horizon.

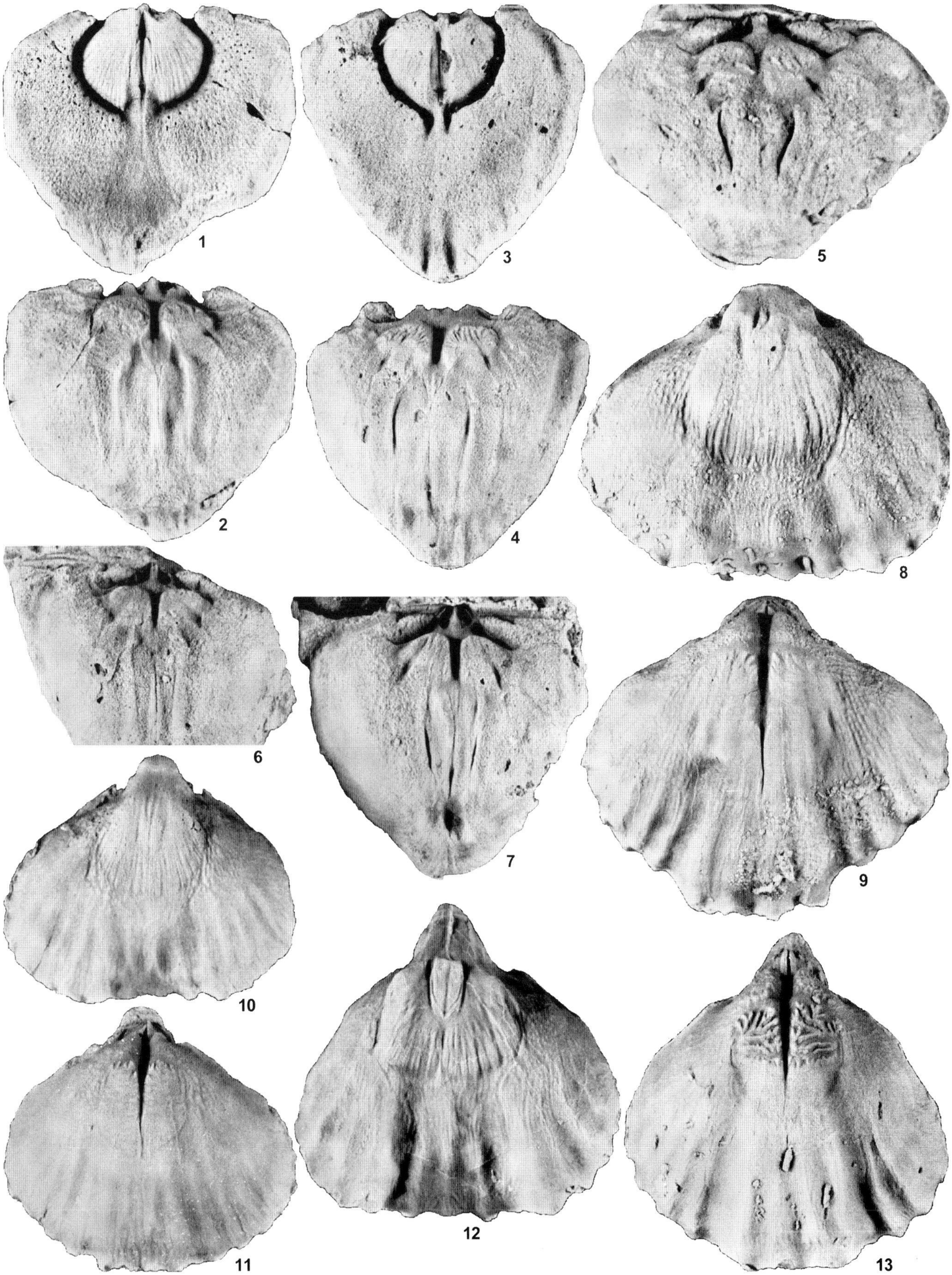
1
3
5
2
4
8
6
7
9
10
12
11
13

Plate 21

Fig. 1. *Hiscobeccus gigas* (Wang, 1949)

Hypotype, GSC 117792, cardinal process, septalium, median septum, and lateral cavities, ×28; loc. O-37130, Gunn Member, Stony Mountain Formation.

Fig. 2. *Dinorthis occidentalis* (Okulitch, 1943)

Hypotype, GSC 117793, section (SM 7-6, 1.2 mm away from the shell posterior end) showing strong cardinal process shaft and swollen myophore, ×21.5; loc. O-37130, Gunn Member, Stony Mountain Formation.

Figs. 3, 4. *Gnamptorhynchos manitobensis* Jin and Zhan, 2000

Paratype, GSC 117743, serial sections (RR 5-8 and RR 5-13, 1.3 and 2.3 mm away from the shell posterior end, respectively) showing incipient cardinal process and strong median septum; fig. 3, ×24.7; fig. 4, ×22.3; Garson Quarry, Selkirk Member, Red River Formation.

Figs. 5, 6. *Nasutimena fluctuosa* (Billings, 1860)

Hypotype, GSC 117794, serial sections (SM 3-6 and SM 3-7, 1.1 and 1.3 mm away from the shell posterior end, respectively) showing cardinal process, socket ridges, hinge teeth, and shell pseudopunctae; fig. 5, ×17; fig. 6, ×33.1; loc. O-37130, Gunn Member, Stony Mountain Formation.

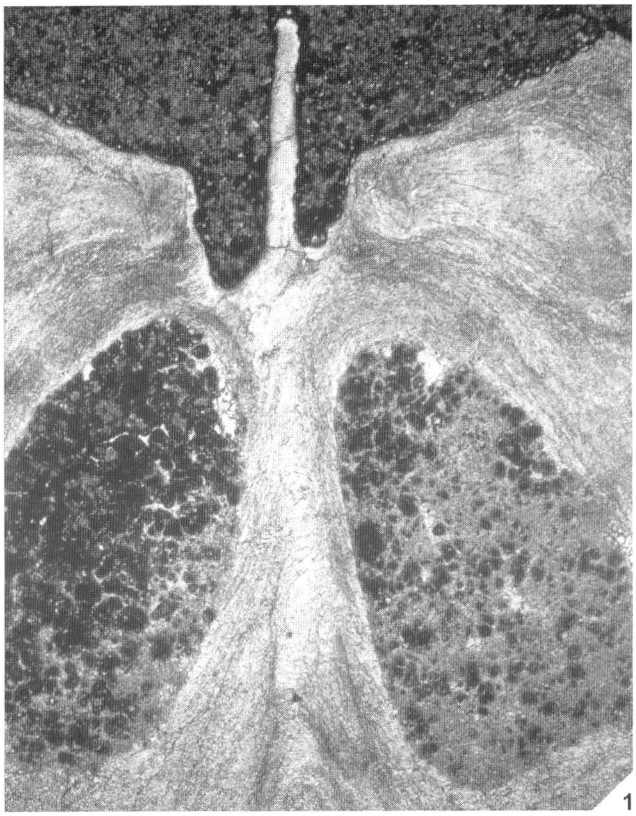
1

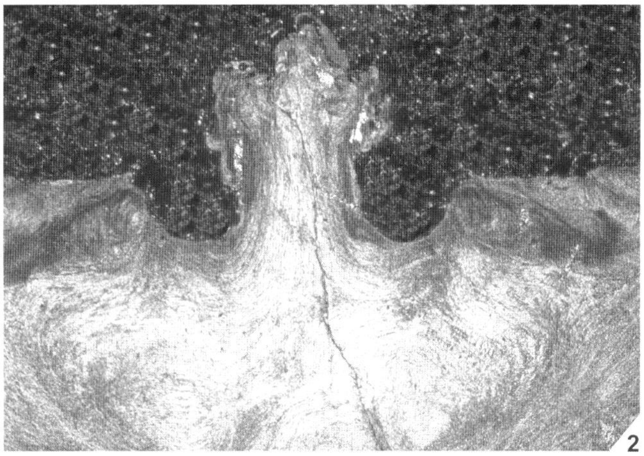
2

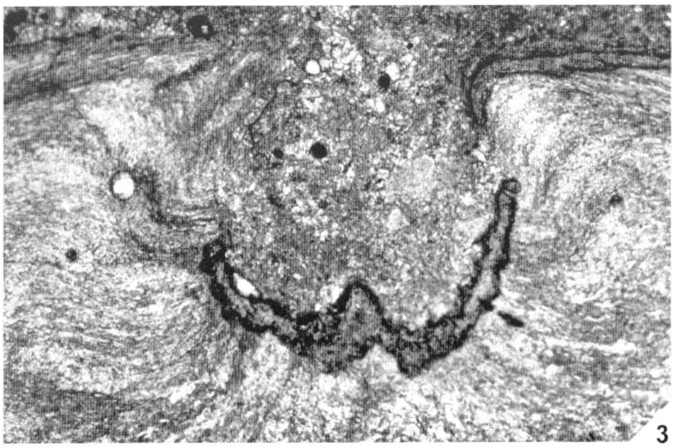
3

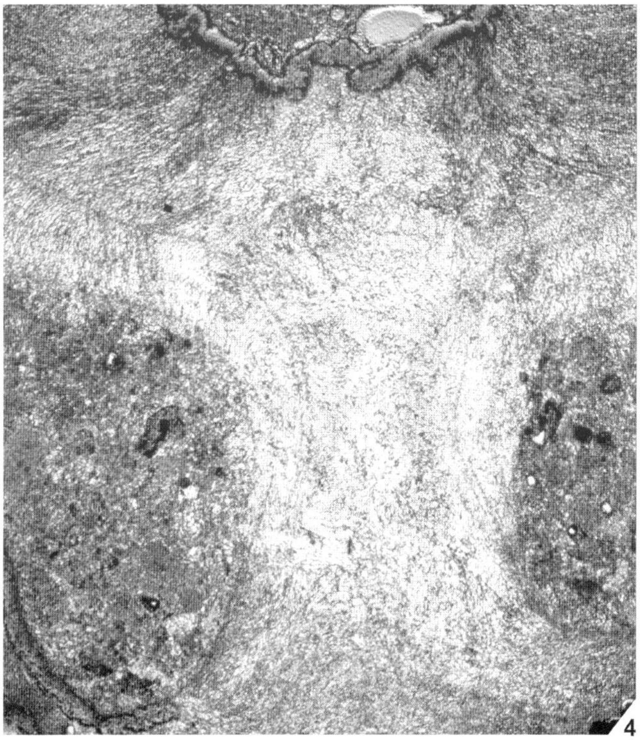
4

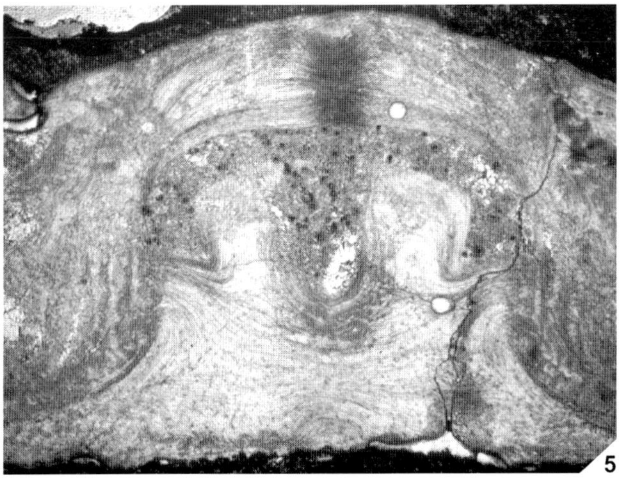
5

6

Plate 22

Figs. 1, 2. *Megamyonia nitens* (Billings, 1860)

Hypotype, GSC 117795, serial sections (SM 5-6 and SM 5-18, 1.2 and 4.6 mm away from the shell posterior end, respectively) showing cardinal process, sockets, socket ridges, and protruding shell pseudopunctae; fig. 1, ×30.7; fig. 2, ×22.1; loc. O-37130, Gunn Member, Stony Mountain Formation.

Figs. 3, 4. *Parastrophinella cirrita* n. sp.

Paratype, GSC 117796, serial sections (RR 3-14 and RR 3-12, 2.2 and 1.8 mm away from the shell posterior end, respectively) showing spondylium, ventral median septum, inner plates, outer plates, alate plates, and fibrous layer of the shell wall; fig. 3, ×32; fig. 4, ×56.4; Garson Quarry, Selkirk Member, Red River Formation.

Figs. 5, 6. *Hypsiptycha anticostiensis* (Billings, 1862)

Hypotype, GSC 117797, serial sections (RR 8-12 and RR 8-8, 2.3 and 1.8 mm away from the shell posterior end, respectively) showing distally inflated cardinal process, median septum, lateral cavities, and deltidial plates; fig. 5, ×38.5; fig. 6, ×23; same locality and horizon.

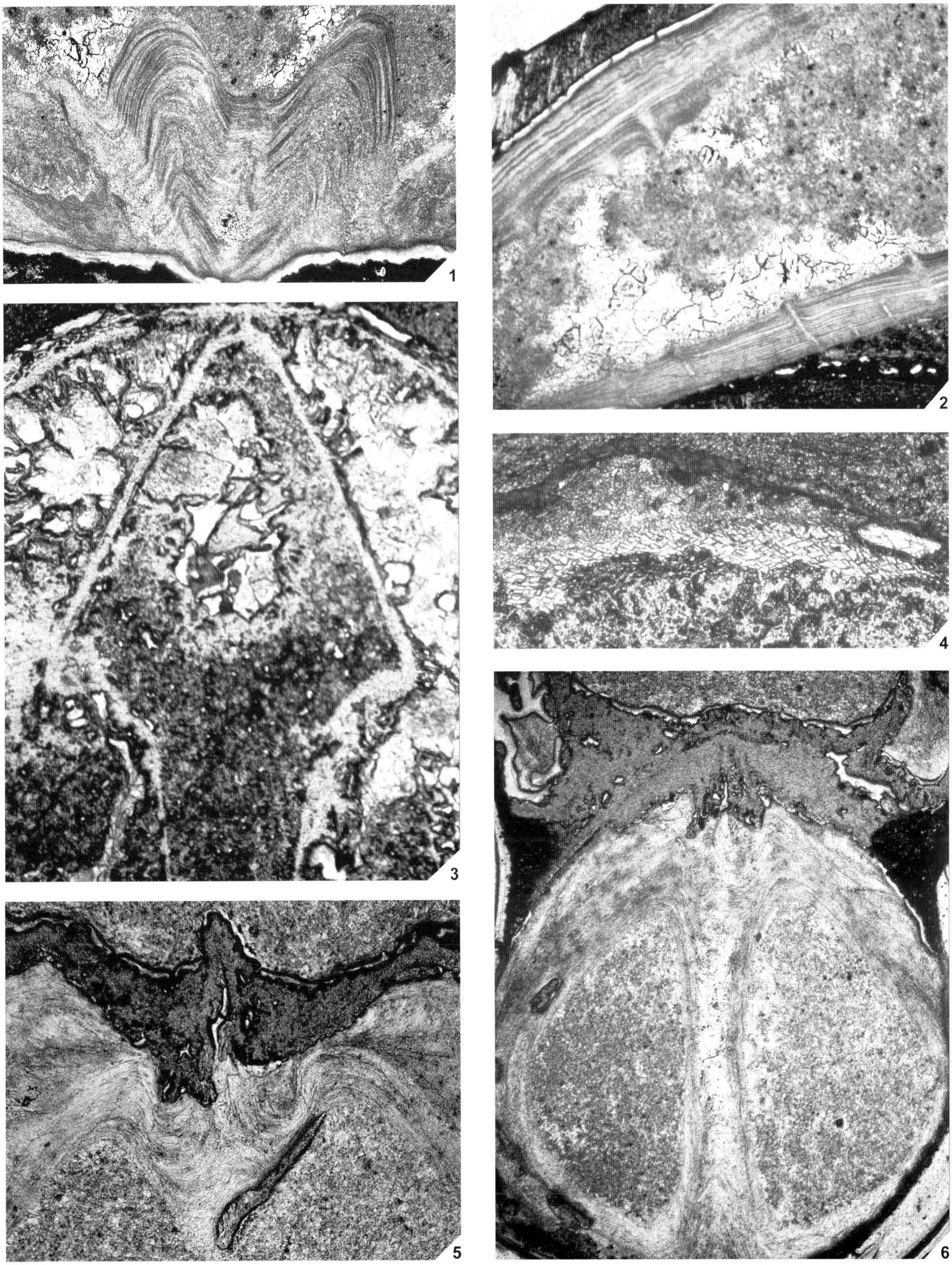
1
2
3
4
5
6

PLATE 23

Fig. 1. *Hypsiptycha anticostiensis* (Billings, 1862)

Hypotype, GSC 117797, cardinal process in the notothyrial cavity, ×48; Garson Quarry, Selkirk Member, Red River Formation.

Figs. 2, 3. *Hiscobeccus capax* (Conrad, 1842)

Hypotype, GSC 117798, serial sections (RR 2-10 and RR 2-18, 1.5 and 2.3 mm away from the shell posterior end, respectively) showing cardinal process, median septum, septalium, and hinge plates; fig. 2, ×24.3; fig. 3, ×34.8; same locality and horizon.

Figs. 4, 5. *Hypsiptycha occidens* (Wilson, 1926)

Hypotype, GSC 117799, serial sections (SM 2-10 and SM 2-15, 1.7 and 2.5 mm away from the shell posterior end, respectively) showing deltidial plates, cardinal process blade-like in the notothyrial cavity but distally inflated, hinge teeth, and socket ridges; fig. 4, ×30.4; fig. 5, ×28.1; GSC loc. O-37130, Gunn Member, Stony Mountain Formation.

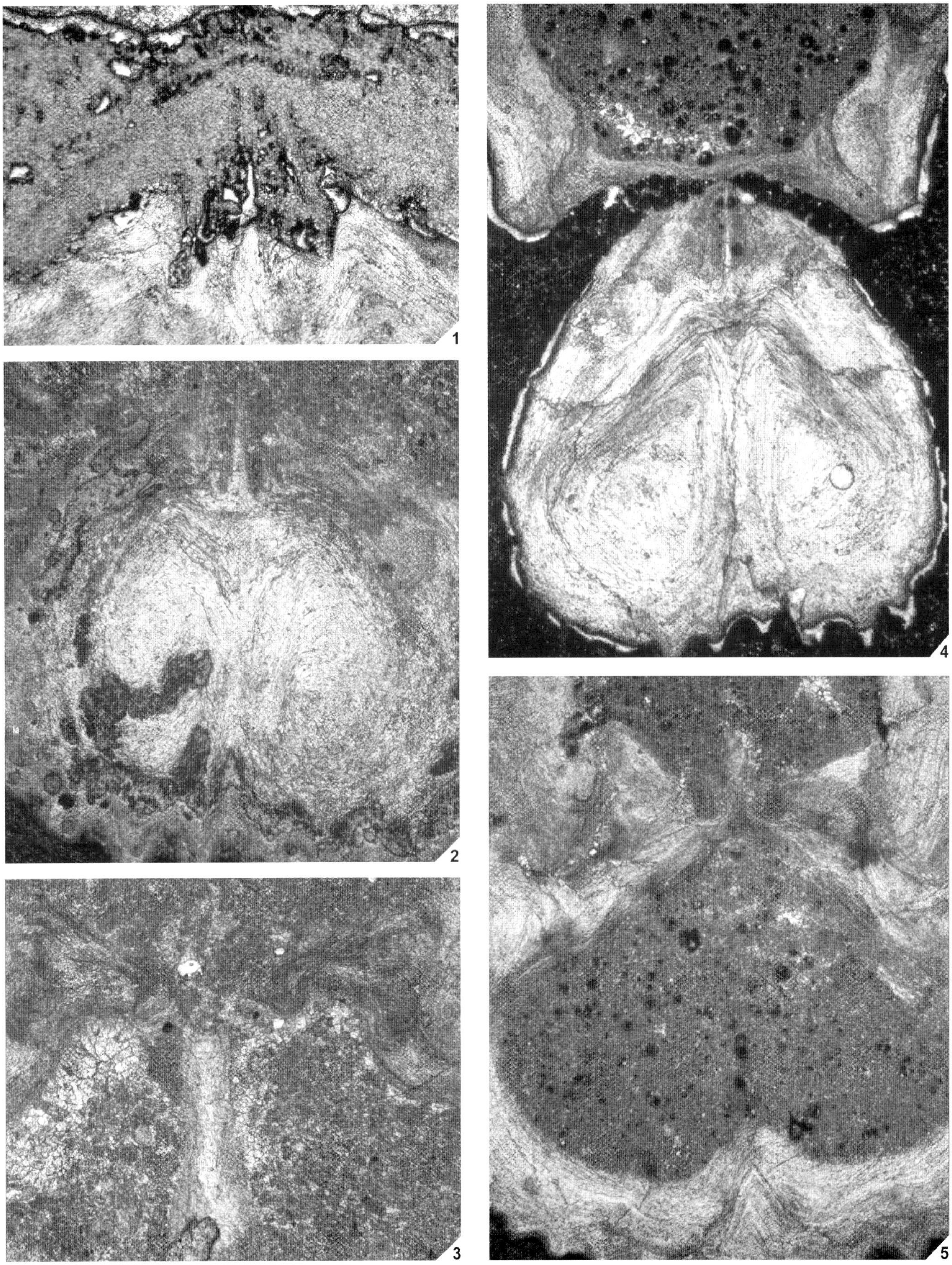
1
2
3
4
5